A RESEARCH HANDBOOK
FOR PATIENT AND PUBLIC INVOLVEMENT RESEARCHERS

Manchester University Press

Copyright © Manchester University Press 2018

While copyright in the volume as a whole is vested in Manchester University Press, copyright in individual chapters belongs to their respective authors.

An electronic version of this book is also available under a Creative Commons (CC-BY-NC-ND) licence, which permits non-commercial use, distribution and reproduction provided the editor(s), chapter author(s) and Manchester University Press are fully cited.

Published by Manchester University Press
Altrincham Street, Manchester M1 7JA
www.manchesteruniversitypress.co.uk

British Library Cataloguing-in-Publication Data
A catalogue record for this book is available from the British Library

ISBN 978 1 5261 3653 4 paperback

ISBN 978 1 5261 3652 7 open access

First published 2018

The publisher has no responsibility for the persistence or accuracy of URLs for any external or third-party internet websites referred to in this book, and does not guarantee that any content on such websites is, or will remain, accurate or appropriate.

The EQUIP project is funded by the National Institute for Health Research's Programme Grants for Applied Research Programme (Grant Reference Number RP-PG-1210-12007) and Greater Manchester Mental Health NHS Foundation Trust (formerly Manchester Mental Health and Social Care Trust (MMHSCT)).

Graphic Design www.greg-whitehead.com
Illustration www.mistermunro.co.uk

A RESEARCH HANDBOOK
FOR PATIENT AND PUBLIC INVOLVEMENT RESEARCHERS

Edited by Penny Bee, Helen Brooks, Patrick Callaghan and Karina Lovell

Manchester University Press

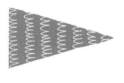

Foreword

Patient and public involvement in research is a requirement of most major health research funders. User-led research and involvement activities are important in shaping and determining research questions, assessing research proposals and guiding and informing research processes. All of these tasks require specific areas of knowledge and skills that can be difficult for members of the general public to acquire.

People who get involved in health research as experts from experience now have a text book to support their research involvement journey. This book will be useful for many, including students at college or university, people working in health research and members of the public getting involved in developing and delivering research studies. I hope this book will encourage members of the public to become involved in health research, and build confidence in their own contributions, because we need them to be involved in research from the very beginning.

The material presented in this book derives from a face-to-face methods course developed for public and patient representatives working on the EQUIP study. It is a book written in partnership between the study academics and the mental health service users or carers who worked with them as advisors and research assistants.

The EQUIP team did not start out with the idea of writing a book. It was the experience of working together over five years that led to this joint venture. If my own experience is anything to go by, patient and public involvement in research is often a reciprocal journey. Academic members of the EQUIP team learnt how to integrate expertise from experience into their work, and the service user and carer representatives positively shaped their research in new and unplanned ways.

And what a brilliant output we have all been given. A very comprehensive book with a topic list, covering all the basics needed for large-scale health research projects: systematic reviews; research design and analysis using both qualitative and quantitative approaches; health economics; research ethics; impact and dissemination. The book is well written and interesting, with a mix of research practices clearly outlined, and an insight into how public and patient representatives can be involved in them and shape decisions. We learn how a carer worked with the team to author a peer review paper and a service user co-delivered a training intervention. We learn how a service user and carer advisory panel influenced study outcome measurement and how service users and carers become involved in focus group data collection and dissemination.

I run a charity that champions the involvement of experts by experience in mental health services research projects. We work with people to create lived experience advisory panels and to support peer researchers delivering user-led projects or collaborative studies. This book is going on the reading list for all our staff, and for the service user and carer advisors we work with, as part of their induction process. Thank you to all the EQUIP team for putting time and effort into this co-produced project and for sharing your partnership with us.

Dr Vanessa Pinfold
Co-founder and Research Director, McPin Foundation

Abbreviations

BDI
Beck depression inventory

CBT
Cognitive behavioural therapy

CD-RISC
Connor-Davidson resilience scale

CONSORT
Consolidated standards of reporting trials

COREQ
Consolidated criteria for the reporting of qualitative research

EDI
Eating disorder inventory

EQUIP
Enhancing the quality of user involvement in mental health care planning

ESRC
Economic and social research council

FGA
First generation antipsychotic medication

ICER
Incremental cost-effectiveness ratio

NHS
National health service

NICE
National institute for health and clinical excellence

NIHR
National Institute of Health Research

OECD
Organisation of economic co-operation and development

PANSS
Positive and negative syndrome scale

PICO
Population, intervention, comparison, outcomes

PICo
Population, interest, context

PPI
Patient and public involvement

PREM
Patient reported experience measure

PROM
Patient reported outcome measure

QALY
Quality-adjusted life year

REC
Research ethics committee

RCT
Randomised controlled trial

SAP
Statistical analysis plan

SD
Standard deviation

SGA
Second generation antipsychotic medication

SOAS-R
Staff observation aggression scale - revised

SUCAG
Service user and carer advisory group

WTPT
Willingness to pay threshold

HOW THIS BOOK CAME ABOUT

This book was developed during a five-year research programme funded by the UK's National Institute for Health Research (NIHR). This study aimed to improve service user and carer involvement in care planning in mental health services. The study was called Enhancing the Quality of User Involved Care Planning in Mental Health Services (EQUIP).

As part of our work on EQUIP, we developed and delivered a successful research methods course for service users and carers. The aim of this course was to help these individuals engage with our research and research team and to work together in true partnership.

This book has arisen out of our partnership, and has been co-written with our service users and carers.

Its aim is to help other public and patient representatives increase their understanding and skills in research methods.

The EQUIP programme used a range of different research methods to achieve its goals, and you will read more about these as you progress through this book. More detail on the EQUIP programme is provided on page 8.

The EQUIP programme involved patient and public representatives with lived experience of mental health services, but all of the research methods that we discuss are used in both physical and mental health research. Whatever your background, or health experiences, this handbook could be helpful.

HOW THIS BOOK IS PRESENTED

Each chapter will provide a brief overview and outline key learning objectives before moving onto the main body of the chapter. All chapters end with a reflective exercise, to help you check what you have learnt. Where helpful, there are also some suggested sources of additional reading.

You will find stories from some of the PPI representatives who worked on EQUIP scattered throughout. These stories reflect on our representatives' own experiences of being involved in the different types of research we discuss.

Andy, Lindsey, Lauren, Debbie and Joe attended our first research methods course. Andy, Lindsey, Lauren joined the EQUIP team as grant co-applicants and researchers. Along with Garry, Debbie and Joe became members of our Service User and Advisory Panel.

Andy, Lauren, Debbie and Garry have lived experience of mental health difficulties, and of using mental health services. Lindsey is a carer for her son, who lives with psychosis. Joe has worked as a mentor with AnxietyUK. We hope you find their stories interesting.

An Overview of the EQUIP Study

The EQUIP study aimed to improve service user and carer involvement in care planning in mental health services. We co-developed with service users and carers a training package for mental health professionals so that they would be better equipped to involve users and carers in their care. Service users and carers helped to design, shape and conduct the EQUIP study and you can learn more about their experiences in Chapter 1.

The content of the training was 'evidence-based.' This means that it was built upon detailed knowledge of the care-planning process, why service user and carer involvement in care planning may not have happened in the past, and what might be the best way of making sure it happens in the future. You will learn more about how to find and use research knowledge and evidence in Chapter 2.

Training was delivered by service users and carers, researchers and health professionals to a range of mental health workers including doctors, nurses, social workers and occupational therapists working in Community Mental Health Teams. We examined whether our new training course led to health professionals involving users and carers in their care. We also looked at how this training influenced health service delivery costs. To do this we used a specific research design, called a randomized controlled trial. You can learn more about this and other research designs in Chapter 3. You can learn more about the type of data we can collect in a trial and how to analyse this in Chapters 4 and 5.

During the EQUIP programme, we worked with service users and carers to develop a new instrument, a patient-reported outcome measure, to measure the extent to which people were involved in their own care planning. You can read more about the importance of patient-reported outcome measures and how you might design and test them in Chapter 6.

We explored the organisational changes that needed to be made by Community Mental Health Teams and the wider healthcare system to improve user and carer involved care planning. We did this by talking to different people and listening to their different views and perspectives. You can learn more about this research approach in Chapters 7 and 8.

We conducted all our work according to the principles of ethical research, and these are discussed more fully in Chapter 9.

Finally, we used lots of different ways to tell users, carers, health professionals and managers about our research findings, especially how user and carer involvement in care planning could be improved. You can learn more about different ways to disseminate research in the final chapter of this book, Chapter 10.

CHAPTER 1:
PATIENT AND PUBLIC INVOLVEMENT (PPI) AND THE RESEARCH PROCESS

Andrew C Grundy

CHAPTER OVERVIEW

This chapter defines and introduces the different stages of the research process: from identifying a problem, to reviewing the literature; then developing a research question; designing a study; obtaining funding and ethical approval; recruiting participants; collecting and analysing data; and reporting and disseminating findings. This chapter will outline how users of health services, their carers and family members, and other members of the public can be involved in these different research stages, and demonstrate the impact that this involvement can have. Examples of different ways of involving and engaging public members in research studies are drawn from the Enhancing the Quality of User-Involved Care Planning in Mental Health Services (EQUIP) research programme.

LEARNING OBJECTIVES

By the end of this chapter you should be able to:

1. Understand the different stages of the research process
2. Understand the impact of Patient and Public Involvement in research
3. Understand the different ways you could be involved

INTRODUCTION

The Frascati Manual provides an internationally recognised definition of research. It defines research as:

> *CREATIVE WORK UNDERTAKEN ON A SYSTEMATIC BASIS IN ORDER TO INCREASE THE STOCK OF KNOWLEDGE INCLUDING KNOWLEDGE OF MAN, CULTURE AND SOCIETY, AND THE USE OF THIS STOCK OF KNOWLEDGE TO DEVISE NEW APPLICATIONS.*
>
> (OECD, 2002, pp30)

Inviting members of the public to offer a lay, non-specialist perspective on the design and conduct of research studies is typically referred to as 'Patient and Public Involvement' (PPI). PPI is a term used to denote meaningful involvement in the design and conduct of a research study. It does not mean a person is involved in a research study as a participant.

At the very least, a research study should be able to evidence consultation with service users, and at best, collaboration and partnership with them as an equal and valuable part of the research team.

Meaningful involvement is not always easy to achieve. One of the most important factors influencing the outcome of involvement is the perspective of different team members, and the different skills, assumptions, values and priorities that each of them brings. Acknowledging and working with these different perspectives is precisely what makes PPI so valuable, but it can also be what limits its success. Effective PPI requires that equal respect is afforded to academic and patient and public researchers, that the perspectives of both parties are equally valued, and that the team as a whole develops and maintains a shared language and goal. It is also important that PPI opportunities are advertised as widely as possible to ensure that they are accessible to a broad range of people from different backgrounds. Adequate training should be provided to ensure that people can be involved in research in a meaningful way (e.g. research methods and how clinical services are organised and commissioned).

Figure 1 **Levels of service user involvement**

Since the mid-1990s, increasing emphasis has been placed on the importance of PPI. The desire to strengthen the involvement and engagement of service users, carers and members of the public in research has been driven by:

a. a strong moral argument that any publicly funded research that aims to benefit health status or health services should be shaped and informed by the people it will affect (Hanley, 2012)

b. accumulating evidence of the benefits of patient and public involvement in research (Staley, 2015)

c. recognition that service users and carers, by virtue of their lived experience, can bring a wealth of experiential knowledge and expertise to the design and conduct of research studies (Faulkner, 1997; Repper, 2008).

Different roles and opportunities for patient and public members to participate in research have emerged. Some people may wish to act as consultants, advising on multiple projects during the early phases of research commissioning and design. Others may choose to engage in one specific project, joining a project advisory panel who will guide and advise a research team throughout the life of that project. PPI advisory panels (sometimes called a Service User and Carer Advisory Groups (SUCAG) or Expert User Groups) provide an independent viewpoint on research progress, advising on research procedures and challenges as they arise and assisting with dissemination.

Service users and carers may also choose to be researchers. Unlike advisory panel members, these individuals work as trained, integral members of the research team, contributing to the design and conduct of the study, and in some cases, its funding application. Service users and carers can be named as principal investigators or study co-applicants. As principal investigators they might take the lead in managing, designing and carrying out a study, or in forming a collaborative team, in what is sometimes called user-led or user-controlled research.

The EQUIP study was conducted by a mix of researchers from different backgrounds. The different contributions that service users and carers made to the EQUIP research programme are shown in Figure 1. As you can see, service users and carers worked in many different roles and had a range of different experiences, and you will learn more about their personal stories throughout the book. First though, let's take a journey through the research process and look more closely at where and when PPI opportunities can arise.

Beginning the research process

The beginning of the research process involves identifying a problem, reviewing the current literature to clarify what is already known and developing a research question to resolve remaining uncertainties or fill knowledge gaps. This is rarely a straightforward process.

Reviewing the literature on a particular topic can help to identify relevant papers quickly, enabling researchers to build upon, rather than duplicate, existing work. It can help to narrow a broad problem down to a specific issue, assess its importance and develop an appropriate and meaningful research question. We will learn more about how to conduct a literature review in Chapter 2.

It is good practice to involve stakeholders in the review process (Rees and Oliver, 2012). In research, the term 'stakeholder' is often used to refer to those individuals, groups or communities who have an interest in, and are likely to be affected by, the conduct and findings of a research project. In mental health research for example, important stakeholders can include service users, carers, wider family members, mental health professionals, service managers and commissioners. Working with stakeholders to define the appropriate focus for a literature review, and identifying and prioritising the research questions that might arise from it, is therefore an important step in making sure any future studies have relevance and applicability to health services (Arksey and O'Malley, 2005). We know that people who use mental health services, carers and professionals have different views about effective care, with professionals often prioritising a clinical model of care, and service users emphasising a social model of care (Rose, 2003). Similarly, they may also have different research priorities.

Service users want research that makes a noticeable difference to their care experiences, both personally and generally (Beresford, 2005). More importantly, they want research that leads to positive improvements in the whole of people's lives, not just in the design and delivery of mental health services (Faulkner and Layzell, 2000). Given these priorities, it's crucial that service users, carers and public members are consulted, or even better, are asked to collaborate in the early stages of the research process, helping to prioritise research ideas and to frame research questions. Involving service users should lead to questions that are more relevant and meaningful to participants. Where service users are not part of the actual study team, this kind of involvement can be achieved through holding focus groups, discussions or local and national stakeholder events.

Designing the study

Having developed a research question, it is important to decide which methods might be the best to answer it. Research generally falls into two types: quantitative research and qualitative research. Quantitative generates numerical data, often through the use of large studies, using methods such as questionnaires and surveys. We will learn more about how to collect and analyse quantitative data in Chapters 3 and 4. Qualitative research explores attitudes, behaviour and experiences through methods such as interviews, focus groups or observation, and we will learn more about this in Chapter 7 and 8.

Once the type of research is set, the study needs to be designed in detail. Collaborating with service users and carers in the design of a research study allows researchers to understand how best to approach potential research participants, why people might drop out of research studies (Goward et al., 2006), why an intervention might work from the user/carer perspective (Allam et al., 2004), what people might find most useful about different interventions and what might be the most appropriate outcomes to measure (Faulkner, 1997).

When evaluating if, and how well, a new intervention works for instance, it is often necessary to ask people to report treatment 'outcomes.' Popular outcomes for mental health interventions might include scales that measures symptoms, recovery, hope or daily activities and functioning. Treatment costs or the need to use other services might also be measured. Interestingly, the most common clinical measures are often the ones that service users like the least, because they do not tap into the priorities of service users themselves (Crawford, 2011). Service users want measures that can capture both the negative and also the positive effects of treatments, and are often willing to complete longer questionnaires to ensure that this is possible (Kabir and Wykes, 2010).

The act of completing questionnaires for service users and carers can in itself be challenging. Collaborating with service users and carers to select and prioritise outcome measures for quantitative research studies is therefore incredibly important, and can help to minimise the number of questionnaires or questions that participants miss out or refuse to answer. It can also help to get feedback on the length of time needed to complete any questionnaires.

Increasingly, service user involvement is also being sought in the design of new outcome measures, for example in selecting possible questions, prioritising questions and/or reducing the number of questions included in a scale, and commenting on ease of response and the emotional impact of its wording (Wiering et al., 2016). Is it too distressing or demoralising for example? Questionnaire development can be a complex and time-consuming process, and we will learn more about this in Chapter 6.

Similarly, when it comes to designing focus group or interview topic guides, service users often ask different questions to non-service users (Rose et al., 2004). Gillard et al. (2010) compared 'academic-researcher' and 'service user-researcher' questions and found that the latter were more concerned with 'how things felt' rather than 'what happened next'. They may also ask questions in a different way, using different phrases and words. It is therefore crucial to involve service users, carers and public members in this stage of the research process.

Funding

Depending on their size and purpose, research studies can be expensive. New research proposals will therefore usually be submitted to a funding body. Funders will look to see if the proposed research study is important (from the funder's point of view); that the proposed methods will answer the research question; that the study represents good value for money; that it be conducted safely and in line with ethical guidelines (see chapter 9); and that the research team are the right people to do the work (Aldridge and Derrington, 2012).

Funders will also want to see that the proposal is well structured and is written simply and clearly, including a summary of the proposed research which is accessible and understandable to members of the public (Aldridge and Derrington, 2012). Service user, carer and public member involvement has an obvious role to play in this, and most funding bodies now mandate PPI in the development of research proposals and grant applications. Many funders also seek to actively include service user, carer and public members in the appraisal of funding applications, both as peer reviewers and as panel members participating in the meetings where funding decisions are finally made.

Figure 2
WHAT ARE FUNDERS LOOKING FOR IN A RESEARCH APPLICATION?
Aldridge and Derrington, 2012

- ✔ The research is important (from funder's point of view)
- ✔ The research will answer the question
- ✔ The proposal represents good value for money
- ✔ The proposal is ethically sound
- ✔ The research team are the right people to do the research
- ✔ A well-structured and well written application
- ✔ A clearly written proposal
- ✔ Includes a lay summary
- ✔ Incorporates patient and public involvement

Ethics

All research studies, with the exception of service evaluations and audits, need to be approved by a Research Ethics Committee (REC) before they can begin. A detailed protocol, which outlines how researchers will deal with any ethical issues (e.g. confidentiality, informed consent etc.), is submitted electronically to the REC and reviewed by a multi-disciplinary team. Researchers may need to attend an REC, if invited, to discuss their application. Additional permissions to carry out research in specific organisations may also be required. We will learn more about research ethics and governance in Chapter 9.

Informed consent is an important principle of ethical research. This means that all potential participants must be allowed to choose to take part in a research study, without fear of losing care, or worrying about what might happened if they don't. To make sure this decision is an informed one, all potential research participants must be given clear and accurate information about why the study is being conducted and what participation would involve. Service user involvement can help to ensure this information is presented clearly and provides all the details that people might want to know. It can ensure that consent is truly informed by making sure that the right information is accessible (Allam et al., 2004) and that potentially offensive, dismissive or misleading statements are avoided (Rose, 2003). Service users may have different perspectives on what might cause distress and how that should be managed (Nicholls et al., 2003).

Figure 3
PPI INVOLVEMENT IN THE DESIGN OF AN ETHICALLY APPROPRIATE STUDY

- ✔ Making information assessable to enable truly informed consent (Allam et al., 2004)
- ✔ Identify potential offensive, dismissive or misleading questions (Rose, 2003)
- ✔ Insight into what might cause distress and how to manage this appropriately (Nicholls et al., 2003)
- ✔ Feedback on proposal and all supporting documentation (e.g. participant information sheets, consent forms, adverts etc.)

Once a study has all the necessary approvals, it can begin to recruit research participants. The precise recruitment strategy that is used will have been outlined in the ethics submission. It could, for example, include recruiting in person at clinics, or via poster display, or via social media. Patient and public representatives who sit on advisory panels can often advise on or act as a conduit to service user networks, potentially increasing access to people who may or may not be in contact with statutory health services. In some instances, the research topic will be sensitive, and in these cases, trained service user researchers can play a valuable role in study recruitment. People from 'seldom heard' or marginalised groups may be more willing to participate in a project involving someone they know (Fleischmann and Wigmore, 2000; Ennis and Wykes, 2013).

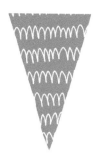

Exactly how data is collected will depend upon the methods chosen to answer the research question. As an integral part of a multi-disciplinary research team, service users, carers and public members can be co-investigators with an active role in collecting data (Hanley, 2012). For example, they could help assist someone in the completion of a questionnaire, or they could facilitate a focus group or face-to-face interview. Often this can enhance the richness and relevance of the data that is obtained. Participants may also choose to share a different type or level of information with someone who has had a similar life story or experience compared to somebody who has not.

Figure 4

Analysing data

The move from collecting to analysing data is rarely a linear process, and exactly how and when researchers begin their analysis will largely depend upon their underlying approach. Quantitative studies will involve data-input and some kind of statistical analysis (see Chapter 4). Qualitative studies may involve identifying themes (or codes) from interview and/or focus group transcripts or observational field notes (see Chapter 8). There is evidence to suggest that service users, carers and public members interrogate qualitative data differently, asking different questions of it and interpreting qualitative data in ways that reflect their priorities (Gillard et al., 2010). These interpretations can be fruitfully pooled with other (non-service user) perspectives to provide a more holistic and meaningful analysis.

Reporting findings

Researchers often talk about 'dissemination' or 'knowledge transfer.' These terms are used to refer to the mechanisms and strategies by which other groups and communities become aware of, obtain and/or subsequently make use of new research findings (Freemantle and Watt, 1994). Service users may have very different dissemination priorities to academic-researchers, whose main emphasis is often upon publication in peer-reviewed journals or presentations at academic and professional conferences. Service users, carers and public members may publish or present the research in their local communities or groups, ensuring wider reach and understanding of the study and its findings. Service users play an important role in framing research findings, in deciding what implications they may have for practice (Hanley, 2012) and in preparing accessible summaries of projects for dissemination that explain the results in a clear and jargon-free way. Increasingly, academic and professional journals are recognising the value of this input and seeking articles led by or co-authored with service users and carers.

Supporting and equipping people to be involved in research

If you are a patient or a carer, somebody who has used health services in the past or a general member of the public, then you will most likely have a unique viewpoint on health service delivery and research, and a valuable contribution that you can make.

Getting involved in research can be extremely rewarding but it is not always easy to do. Barriers include a lack of awareness about opportunities to get involved in research, language barriers, physical or emotional health, a lack of confidence, the behaviour or attitudes of researchers, or inappropriate timing/location of meetings. Research teams need to make sure that people are and can be involved in a meaningful way. This means making sure that PPI members are properly and regularly supported, and that they are fully recognised for their time and knowledge contributions, as important and integral members of the research team. The EQUIP team has found that offering service users and carers a short course in Research Methods and Design can help to facilitate collaborative relationships and can give members the confidence to play their role in the multi-disciplinary research team.

PPI stories from EQUIP

Next, Andrew and Garry (members of the EQUIP team) reflect on their experiences of being involved in the EQUIP programme of research.

Andrew's story

Following the first research methods training course I was invited to be a coapplicant on the EQUIP research programme. This meant that I became a member of the research team for the whole of the EQUIP programme, and shared responsibility for the design, management and conduct of the EQUIP studies.

One of my roles was to promote the studies across Nottinghamshire, as this was one of our main research sites. It's been a real privilege to network with local service user and carer groups, who have invited me back time and time again to talk about our work. I've really appreciated their help and support!

I received some practical training in qualitative research methods, and thoroughly enjoyed co-facilitating focus groups and conducting interviews in the EQUIP programme. It was wonderful to take a lead on the analysis of the service user data, to help write up our findings, and to assist in developing an animation to outline our new model of care planning involvement. It was also really interesting to see how new questionnaires are developed and tested – and it's been great to be part of producing a new measure of service user involvement in care planning, and an audit tool that services can use to improve their care.

Co-delivering the EQUIP training intervention took up a lot of my time, but I absolutely loved it. I was involved in recruiting mental health teams to the trial, in co-facilitating their two days of training and in providing subsequent clinical supervision. It's been a privilege to help Trusts consider how best to implement our training, working with service users, carers and clinicians so that they can go on to deliver the training themselves.

It's been a joy to be able to present at events and conferences, I've loved telling people what we've been finding out. I've had a few mental health 'blips' along the way, but I feel that with the right training and mentorship I've been able to use my lived experience and expertise in a constructive way.

Garry's story

During the time I've been involved in EQUIP, I've really valued the meetings that have taken place. As a group we've dedicated time to looking at the process of care planning. In the focus groups that I have observed, service users have been very open and very honest, sharing experiences in a safe and supportive setting.

We asked people to talk about their positive experiences, what has worked well for them and also what could be improved. We gathered a great deal of information that has been to the EQUIP programme and has been published as important research findings. I think focus groups are very good; by helping people to share their feelings and experiences, we can travel a long way towards improving the care that we receive.

 ## Reflective exercise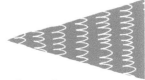

- What opportunities are there for members of the public to get involved in health services research?
- Are there any stages of the research process that are of particular interest to you? Why?
- What skills can you offer a research team?
- How might the chapters in this book help you and what other training might you need?

REFERENCES AND FURTHER READING

1) Aldridge, J. and Derrington, A. M. (2012) The Research Funding Toolkit: How to Plan and Write Successful Grant Applications. London: Sage.

2) Allam S., Blyth S., Newman, A. and Repper, J. (2004) Researching collaboratively with service users, Journal of Psychiatric and Mental Health Nursing 11:3, 368-373.

3) Arksey, H. and O'Malley, L. (2005) Scoping studies: Towards a methodological framework, International Journal of Social Research Methodology: Theory & Practice 8:1, 19-32.

4) Beresford, P. (2005) Theory and practice of user involvement in research: making the connection with public policy and practice. Chapter 1 in L. Lowes and I. Hulatt (Eds.) Involving service users in health and social care research. London: Routledge.

5) Crawford, M.J., Robotham D., Thana, L, Patterson, S., Weaver, T., Barber, R., Wykes, T. and Rose, D. (2011) Selecting outcome measures in mental health: the views of service users. Journal of Mental Health 20:4, 336-46.

6) Ennis, L., and Wykes, T. (2013) Impact of patient involvement in mental health research: longitudinal study. British Journal of Psychiatry 203:5, 381-386.

7) Faulkner, A., (1997) Knowing our own minds: Users views of Alternative and Complementary Treatments in Mental Health. London: Mental Health Foundation.

8) Faulkner, A. and Layzell, S. (2000) Strategies for Living: A report of user-led research into people's strategies for living with mental distress. London: Mental Health Foundation.

9) Fleischmann P. and Wigmore J. (2000) Nowhere else to go: Increasing choice and control within supported housing for homeless people with mental health problems. London: Single Homeless Project.

10) Freemantle, N. and Watt, I. (1994) Dissemination: implementing the findings of research. Health Information and Libraries Journal 11:2, 133-137.

11) Gillard, S., Borschmann, R., Turner, K., Goodrich-Purnell, N., Lovell, K. and Chambers, M. (2010) 'What difference does it make?' Finding evidence of the impact of mental health service user researchers on research into the experiences of detained psychiatric patients. Health Expectations 13:2, 185-194.

12) Goward, P., Repper, J., Appleton, L. and Hagan, T. (2006) Crossing boundaries: identifying and meeting the mental health needs of gypsy travellers. Journal of Mental Health 15:3, 315-327.

13) Hanley B. (2012) Involving the public in NHS, public health, and social care research: Briefing notes for researchers. Eastleigh: INVOLVE.

14) Kabir, T. and Wykes T. (2010) Measures of outcomes that are valued by service users., Chapter 1 in G. Thornicroft and M. Tansella (Eds) Mental Health Outcome Measures. RCPsych 3rd edition.

15) Nicholls, V., Wright S., Waters, R. and Wells, S. (2003) Surviving user led research: Reflections on supporting user led research projects. London: Mental Health Foundation.

16) Organisation for Economic Cooperation and Development [OECD] (2015) The Measurement of Scientific and Technological and Innovation Activities, The Frascati Manual: Guidelines for Collecting and Reporting Data on Research and Experimental Development. OECD Publishing: Paris.

17) Rees, R. and Oliver, S. (2012) Stakeholder perspectives and participation in reviews., Chapter 2 in D. Gough, S. Oliver and J. Thomas, An Introduction to Systematic Reviews. London: Sage.

18) Repper, J. (2008) Carers as Researchers: Lessons from the Partnerships in Carer Assessment Project'. In M. Nolan, E. Hanson, G. Grant, and J. Keady (Eds) User Participation in Health and Social Care Research, Open University Press.

19) Rose D. (2003) Collaborative research between users and professionals: peaks and pitfalls. Psychiatric Bulletin 27, 404-406.

20) Rose D., Fleischmann, P. and Wykes T. (2004) Consumers' perspectives on ECT: A qualitative analysis. Journal of Mental Health 13:3, 285-294.

21) Staley, K. (2012) An evaluation of service user involvement in studies adopted by the Mental Health Research Network. London, MHRN.

22) Thornicroft, G., Rose, D., Huxley, P., Dale, G. and Wykes, T. (2002) What are the research priorities of mental health service users? Journal of Mental Health 11:1, 1-5.

23) Wiering, B., de Boer, D., and Delnoij, D. (2016) Patient involvement in the development of **patient-reported outcome measures: a scoping review.** Health Expectations 20, 11-23.

Chapter 2:
Introduction to Systematic Reviews

Kelly Rushton and Owen Price

Chapter Overview

Health professionals must make sure that the treatments they provide are safe and have been proven to work. This means they must be able to access and understand all the available research relating to treatments that they use. There can be hundreds of research studies conducted on each treatment, so it is often not practical for busy health professionals to gather all of the evidence themselves.

Deciding on a treatment on the basis of a single study or a select number of studies is not recommended. Individual studies may have been poorly conducted and have misleading results, or the findings of the study may conflict with other studies that were not accessed by the health professional.

This is why it is important that researchers gather all of the available evidence relating to each treatment, analyse and combine that evidence, and make it available to healthcare professionals in a comprehensible way. This allows professionals, and their patients, to make accurate judgements about a treatment's risks and benefits.

The method used by researchers to ensure that this process of identifying, analysing and combining the findings of multiple studies is sufficiently rigorous, is called a systematic review. This chapter will detail the steps involved in a well-conducted systematic review.

LEARNING OBJECTIVES

By the end of this chapter you should be able to:

1. Understand what systematic reviews are
2. Understand the stages involved in undertaking a systematic review
3. Understand why systematic reviews are important

INTRODUCTION

Traditionally, systematic reviews have combined evidence from treatment trials and have answered 'yes/no' questions such as 'Does this treatment work?', 'Is it safe?' This involves combining 'quantitative' data (data related to numbers) across a large number of participants and studies and calculating what the average benefit (or risk) of a treatment might be. More recently, systematic reviews have also combined findings from 'qualitative' studies. Qualitative studies explore people's views, experiences and beliefs in an in-depth way, using, for example, individual or group interviews. This type of evidence is important. For example, understanding the way service users feel about a service or treatment can provide important indications of how likely they are to continue to use it. Systematic reviews based on qualitative data can answer important questions about how, why and when a service or treatment might work, rather than 'does it work.' The information this provides can be used to change treatments and services to better meet the needs of service users. Again, a more complete understanding of views and experiences can be generated from combining the findings of many qualitative studies than from relying on the findings of one study alone.

Conducting a review

Irrespective of whether you are conducting a systematic review involving qualitative or quantitative evidence, there are a series of steps that must be followed to ensure that your review is properly conducted. These are as follows:

Defining the question
(what is not already known and what will the review provide an answer to?).

Data synthesis
(the process used to combine and summarise all of the available data).

Searching the literature
(how and where studies will be identified).

Data extraction
(how results will be transferred from the original studies to a review spreadsheet, so that the results of all the included studies can be viewed together).

Quality assessment
(to judge how well the original studies were conducted, and to identify any particular strengths or weaknesses that might have influenced their results).

The following sections will describe these steps in more detail.

Importantly, how each of these stages is conducted should always be agreed and written up in a **review protocol** before starting the systematic review. The review protocol should provide detail on all the stages of the proposed review. This helps to prevent researchers making on-the-spot decisions. It reduces error or 'bias' in the findings and helps others to judge how well the review was completed i.e. did it address its original aims and objectives and follow the right methods?

Defining the review question

Developing a good review question is perhaps the most important step in conducting a systematic review. The question will inform the search methods, guide researchers in deciding which studies should be included and excluded and determine whether or not the review produces findings that are meaningful and useful to health services and their users.

It is really important that the review question is well-constructed. The review question must have potential to generate new knowledge and understanding. Put simply, is the review worth doing? Is the question worth answering?

The review question must be precise enough to ensure that the review can be completed. Systematic reviews are normally undertaken by a team of people, but can sometimes still take one or two years to complete. A review question that would need hundreds of thousands of studies to answer it would not be feasible to complete.

There are some easy ways to ensure that a proper review question is developed. 'Does it work' (or effectiveness) questions use a technique called **PICO** (Population, Intervention, Comparison, Outcomes). This ensures the following categories are specified in the review question:

- **Population:** the patient or service user group that the review is interested in e.g. children, working-age adults, people living with psychosis
- **Intervention:** the treatment or therapy that the review will evaluate e.g. counselling, anti-psychotic medicines, care-planning
- **Comparison:** the treatment or therapy that the intervention will be compared with e.g. education, usual care or 'no treatment'
- **Outcomes:** the outcomes that will be used to evaluate the intervention's effect e.g. mental health symptoms, quality of life, patient satisfaction

Figure 5 provides an example of how these four categories can be used to develop a focused research question.

Figure 5 Developing a review question using PICO

However, not all review questions will be addressing 'does it work' or 'effectiveness' questions. Reviews that are interested in understanding people's experiences will be searching for qualitative evidence. These types of review use a different 'PICO' technique to structure their questions: PICo (Population, Interest, Context). This ensures that the following categories are specified in the review question:

- **Population:** the patient or service user group you are interested in e.g. adults with bi-polar disorder
- **Interest:** the activity or issue you are interested in e.g. care-planning
- **Context:** the setting of interest e.g. community mental health teams

Figure 6 provides an example of how these three categories can be used to develop a focused research question for a review of qualitative evidence.

Figure 6 **Developing a review question using PICo (review of qualitative evidence)**

What do people with anxiety **report about their** experiences of receiving Cognitive Behavioural Therapy for anxiety in the community?

Searching the literature

Once you have your question it is time to see what studies are already out there. To search the literature in a rigorous way you must:

- Know which electronic databases you are going to search (i.e. identify your **data sources**)
- Have a systematic method for searching the databases to make sure no relevant studies are missed (i.e. develop a **search strategy**)

Searching databases will generate many studies that are potentially relevant to your review question. You will need to sift through these studies to identify those that are and discard those that aren't. To ensure that this is done objectively, without bias, each study is assessed against pre-agreed criteria. These are called the review's **inclusion and exclusion criteria**.

Remember, the arrangements for all of these processes should be pre-agreed and written up in your review protocol before your systematic review begins. Changing the protocol after the review begins is possible but is not desirable.

Selecting the data sources – where to search

Systematic reviews are normally conducted by searching online databases. These databases store nearly all of the research conducted in healthcare and their records stretch back many decades. Figure 7 shows some of the databases that are typically used for systematic reviews. You will notice that some of these are quite general and keep records related to many different areas of health research (e.g. MEDline). Others are more subject-specific. PsycInfo, for example, only includes records of mental health and psychological research. Which databases you choose will depend partly on your review question. As a general rule, it is desirable to include all the databases that you think might hold relevant records. Although this will make your review more time-consuming, it will reduce the likelihood of you missing relevant studies and biasing your review.

All of the databases in Figure 7 can be accessed via a simple Google search but they require expensive subscriptions. Most people access them using a University or hospital subscription, so you may need to ask these organisations for a username and password. Although using these databases is not difficult once you've learnt how to use them, they can seem fiddly and frustrating at first. Each has its own quirks in terms of how they work, so it is advisable to enlist the help of someone who is experienced in using them. Hospital and/or University librarians are often happy to book you in for an individual session on request.

Figure 7 **Examples of electronic databases**

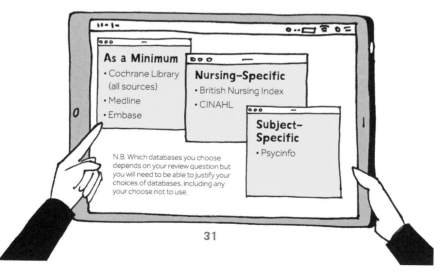

Developing a search strategy

Developing an effective search strategy is vital, because it is this which ensures all relevant studies are included in the review. The term 'search strategy' refers to the terms that are entered into the electronic databases to retrieve studies. It also refers to how these terms are combined to ensure that only studies relevant to the review question are generated by the databases you use. This really is important - if your search generates 20,000 results, you might not have the time check them for relevance. The aim is to capture all relevant results without generating lots of irrelevant results.

A useful technique is to structure your search terms around the PICO categories of your research question. Using the PICO question in Figure 5 (In working-age adults with anxiety, is Cognitive Behavioural Therapy more effective than Citalopram in reducing anxiety symptoms?), you can see that for each PICO component (Population, Intervention, Comparison, Outcome) there are many possible ways of referring to these. For example, Cognitive Behavioural Therapy is often referred to by the acronym 'CBT' and Citalopram can be referred to by its trade name 'Cipramil,' or a more general term such as 'medication.' Of course, different research studies will opt for different terms, so it is important that your search strategy can capture these differences. Figure 8 shows how search terms can be organised around the PICO categories (Population, Intervention, Comparison, Outcomes) using all possible term variants for each category.

Of course, inputting all your search terms into the databases would simply generate a long list of the studies relevant to each term. Really, you want the database to only generate a list of studies that include all of the terms from all your PICO categories (e.g. Population, Intervention, Comparison, Outcome). This will minimise the number of results your search generates, without missing relevant studies. To do this, you have to let the database know which terms are variants within each PICO category, and how the different categories should be combined to generate the best set of results. This is achieved using 'boolean operators.' Boolean operators are simply the words 'AND' or 'OR.' You can enter these after each search word to let the database know whether the word you have used is a variant of other terms in the same category (by using 'OR') or whether you are combining different categories of search terms (by using 'AND.)' Look again at Figure 8 to see how AND and OR have been used to distinguish search term variants and PICO categories.

Figure 8 **Search terms generated using PICO technique**

POPULATION SEARCH TERMS	INTERVENTION SEARCH TERMS	COMPARISON SEARCH TERMS	OUTCOME SEARCH TERMS
Working age **OR** 18-65 **OR** Adult	Cognitive Behavioural Therapy **OR** CBT	Citalopram **OR** Cipramil **OR** Celexa	Anxiety **OR** Worry **OR** Panic

Inclusion and exclusion criteria

Once all your databases have been searched, you will have a long list of studies that are potentially relevant. These need to be sifted to check their relevance and to decide whether or not they will be included in the review. To ensure that this process of sifting and checking is unbiased, it is important that pre-agreed criteria are followed. These criteria are called your inclusion and exclusion criteria. To minimise bias, it is important that two researchers independently check each study against these criteria and compare whether or not they think it should be kept in or out of the review.

The easiest way to develop inclusion and exclusion criteria for a review is to use the PICO framework. Table 1 provides an example of inclusion and exclusion criteria developed for the example review question in Figure 5.

Table 1 **Inclusion and exclusion criteria generated using PICO technique**

PICO component	Inclusion criteria	Exclusion criteria
Population	Participants are working age adults (18-65 years)	Participants are older adults (65+) Participants are children and/or adolescents (<18)
Intervention	Intervention is Cognitive Behavioural Therapy-based	Intervention is a pharmacological treatment Intervention is a non-Cognitive Behaviour Therapy-based psychological therapy
Comparison	Comparison treatment is Citalopram	Comparison treatment is an anti-depressant but not Citalopram Comparison treatment is a psychological therapy
Outcome	Anxiety symptoms are measured as a criterion for intervention success	Outcomes measured relate to depression, no anxiety measure is included

Data extraction

The term 'data extraction' refers to the process of transferring relevant findings from the included studies to a working spreadsheet. This enables the data from all the included studies to be viewed together. Your data extraction spreadsheet might include columns for findings that answer your review question (e.g. study results) as well as columns for summarising important information on study context or quality (for example, the country where the study was conducted, study design, or whether or not the study was conducted in a 'real world' setting). The spreadsheet will have individual rows for each included study.

Quality assessment

Assessing the quality of each study included in a systematic review is important. No research study is perfect. Individual studies may be prone to error or have limitations that influence the findings. If your review does not account for these biases, the results of your review may also be biased.

Broadly speaking, quality assessment draws the reader's attention to important strengths or weaknesses in the research evidence. Sometimes it is used to apportion more or less weight to individual studies in a review, depending upon whether they are judged to be of a higher or lesser quality.

The way in which quality is assessed in a systematic review often depends upon the subject matter and the type of studies being reviewed. Quality checklists can provide a useful way of assessing all of the evidence in a methodical and standardised way. There is not one single accepted tool for quality assessment, but a variety of published quality assessment guidelines and tools are available that might help you in this process.

As a general rule, quality assessment will usually consider:

- How each study was designed and conducted
- How appropriate this design was to fulfil the study's objectives
- How well the intervention was delivered
- How data were measured, collected, reported and analysed
- How well the findings were interpreted
- How relevant or generalizable these results might be to other people outside of the study

Data synthesis method

The term 'data synthesis' refers to the method that is used to combine findings across all of the studies in your review.

The techniques that are used to combine qualitative and quantitative research evidence are very different. For quantitative systematic reviews (those which aim to answer 'does it work' questions), a statistical technique called 'meta-analysis' is often used. Meta-analysis usually combines data from randomised controlled trials that have been carried out to evaluate treatment effectiveness. Numerical data from many different trials are combined, to explore the average benefit of a particular treatment. An example of the type of data used in a meta-analysis might be service users' scores on a depression scale before and after anti-depressant treatment. Meta-analysis is very important because it enables researchers to assess a treatment's benefit in a much larger group of people than could be achieved in a single trial. Broadly speaking, the larger the number of participants, the more confidence you can have in your estimation of a result. Of course, the relevance of your result depends very much on the type of data you combine. Trials conducted in very different groups of people, or with very different interventions, may not be able to be grouped together.

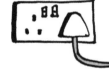

Qualitative systematic reviews group together data, or text, that summarises people's views and experiences. There are many different techniques that can be used to combine this type of data (Barnett-Page and Thomas, 2009), although most share two key stages.

The first stage seeks to summarise the findings of the included studies, by organising them into common themes (often referred to as descriptive themes). The second stage looks more closely at these themes, to try to understand how they relate to each other. This second stage generates 'analytical themes.' Analytical themes provide a higher level of analysis. They extend our understanding beyond the findings of the original studies, and can be a useful way of examining the combined evidence on a topic.

PPI stories from EQUIP

Next Lauren discusses her experiences of being involved in a systematic review.

LAUREN'S STORY

I joined the EQUIP programme at its inception as a participant on a research methods training course for service users and carers. One of the first things we had to do was to decide how we were going to measure service user 'involvement' in care planning. We conducted a systematic review to identify all the research that had already been published on this topic, so that we could understand all the different ways that involvement had been measured in the past. Reviewing was one of the skills that I had learnt on my research methods course.

Taking part in a systematic review can be daunting, but there are many ways that PPI representatives can contribute. With the right training, knowledge and education work we can work as equals and even lead on such work.

We met as a group to plan the review together and to decide exactly what we needed to find out. My first job was to help with the literature search. I helped to develop search terms for the library database searches.

Then I agreed with another researcher which studies should be included in our review and which should be left out. This was a long process but more manageable than I expected. Information, guidance and attention to detail made the task possible. I enjoyed this piece work and was proud of my emerging skills in research. It felt good to have my capabilities recognised.

Next, I had another important role to play – I helped to shape the way that our data was extracted and synthesised. In our review we looked at all the different ways that involvement could be measured. We scored each one according to its quality, i.e. how well it would measure involvement from a researcher's perspective, but we also scored each one according to how easy or hard a service user might find them to complete.

I worked with our service user and carer advisory group (SUCAG) to come up with a 'wish list' for a good involvement measure. Our literature review had found a number of ways of measuring service users' involvement in care decisions, but none of them were acceptable to people with lived experience. PPI involvement had shown us that we needed to develop a new measure!

Like me, you might be asked to shape a review to make sure its results are meaningful and relevant to others. It was great to be able to make sure service users' views were included in our review, and to help write up our findings.

REFLECTIVE EXERCISE

- Why might you carry out a systematic review?
- What are the main stages involved in conducting a systematic review?
- How can the quality of studies in a systematic review be assessed?

ALLIED PAPERS

1) Bee, P., Price, O., Baker, J., and Lovell, K. (2015) Systematic synthesis of barriers and facilitators to service user-led care planning. The British Journal of Psychiatry 207:2, 104-114; DOI: 10.1192/bjp.bp.114.152447.

2) Price, O., Baker, J., Bee, P. and Lovell, K. (2015) Learning and performance outcomes of mental health staff training in de-escalation techniques for the management of violence and aggression. The British Journal of Psychiatry 206:6, 447-455; DOI: 10.1192/bjp.bp.114.144576.

REFERENCES AND FURTHER READING

Barnett-Page, E. and Thomas, J. (2009) Methods for Research Synthesis (Internet). Available from: http://eppi.ioe.ac.uk/cms/Default.aspx?tabid=188

CHAPTER 3:
QUANTITATIVE RESEARCH DESIGN

Owen Price and Karina Lovell

CHAPTER OVERVIEW

Quantitative research uses large samples and, as such, the findings of well-conducted studies can often be generalised to larger populations. However, it is important that studies are well-designed to avoid errors in their interpretation and/or the reporting of inaccurate results. Misleading results from quantitative studies can have serious negative implications such as wasting public money on flawed policies and subjecting service users to ineffective or harmful treatments. This chapter explores descriptive and experimental quantitative research designs and examines, through case examples, the difference between cross-sectional, longitudinal and cohort studies. Factors leading to poorly and well-constructed studies are explored, along with a discussion of the key features of well-designed randomised controlled trials, the gold-standard design for testing treatment effectiveness.

LEARNING OBJECTIVES

By the end of this chapter you should be able to:

1. Explain the importance of well-designed quantitative studies
2. Explore descriptive and experimental quantitative designs
3. Identify the key features of a randomised controlled trial

INTRODUCTION

Quantitative research generally uses large numbers of participants. This is because it seeks to draw accurate conclusions about, for example, how common a health problem is within a population, what factors increase a person's likelihood of developing a health problem (e.g. weight, gender, wealth, employment, education etc.), or whether a medicine or other intervention is effective in treating a health problem. Without larger numbers of people included in studies with these aims, conclusions are less likely to be accurate because they may not reflect how things actually are in the target population. There are two types of quantitative design, descriptive and experimental. This chapter will introduce descriptive and experimental quantitative designs using case examples to bring them to life.

Descriptive quantitative designs

Descriptive studies measure things as they are without intervention from researchers to change people's behaviour and experiences. **Case example A** in **Figure 9** provides an example of a descriptive design:

Figure 9 Case example A Descriptive design

Paul is interested in looking at how common clinical depression is in people living in Bristol and what factors seem to increase the likelihood of people having depression. He sends out a questionnaire to everyone in Bristol measuring their levels of depression and collecting basic demographics such as age, gender, ethnicity, sexuality, accommodation and employment status. Paul hasn't changed anything within the population in order to assess its effect, he's simply described the frequency of depression in that population and some of the characteristics of people with and without depression at a particular point in time.

There are different types of descriptive quantitative research designs. These produce different types of findings, and which type to use depends on the question being asked.

Consider again **case example A**. This type of study is called a **cross-sectional study**, which means it involves researchers taking a snapshot of a population at a single point in time. So **case example A** might give us some useful information about the prevalence of depression in Bristol at that point in time and suggest some possible factors that might increase the likelihood of people having depression. It is a useful design for describing what a situation is, but it is less useful in telling us why the situation is as it is or how stable that situation is over time.

Imagine Paul had distributed his questionnaire to all the people of Bristol once every year from 1997 to 2017. This would be a **longitudinal study**. This design would enable Paul to examine how rates of depression change over time and generate further **hypotheses** (possible explanations to be proven or disproven) about the factors that seem to increase the prevalence of depression in the population or explain differences in the prevalence of depression in different subgroups (e.g. age, gender and ethnic groups). These factors might, for example, include things like changes in mental health policy, funding or service provision during this time. However, a longitudinal study cannot <u>prove</u> relationships between these different factors. It cannot estbalish **causal relationships**. It can only observe possible relationships (called **associations**), i.e. events that seem to happen at the same time. So Paul might be able to say that, based on his longitudinal study, when policy change X occurs, outcome Y also seems to occur. He cannot say conclusively that X has caused Y because there may have been another factor, unknown to Paul, that was actually responsible for the change in rates of depression.

Some descriptive designs can generate stronger indications of relationships between events than others. In a **longitudinal cohort study**, for example, researchers identify two groups (or cohorts) of people that are broadly similar to each other, except for the fact that one group has been exposed to a particular circumstance and the other has not. They then collect data from both groups at multiple time points and examine differences in outcomes between the two. **Case example B** in **Figure 10** provides an example of a longitudinal cohort study.

Figure 10　　Case example B　　Longitudinal cohort design

Debbie is interested in the relationship between children's psychological well-being and grammar school systems. She knows that Trafford borough has a grammar school system and Manchester does not. This provides an opportunity to analyse differences between the two groups as a result of the different school systems the children have experienced. Debbie asks every school-starting child in Trafford and Manchester to complete a questionnaire measuring their psychological well-being and repeats the questionnaire annually throughout each of their school years (4-16).

Debbie wants to ensure, as far as she can, that any differences she observes in psychological well-being between the two groups is a result of differences in the school system they have experienced. Because of this, she identifies further information about each child who completes the questionnaires, including: parental income, education, relationship status and whether English is their first language, the level of social deprivation in the area where the child lives and whether the child has special educational needs. Collecting this information allows Debbie to adjust her analysis to ensure that differences between the two groups (Manchester and Trafford children) other than the school system they have experienced are not accounting for the differences (if any) she observes in psychological well-being.

Debbie has made great efforts to ensure that the differences she observes between the two groups in psychological well-being are a result of the different school systems between Manchester and Trafford. Her study therefore may make a contribution to understanding a possible relationship between different education systems and children's psychological well-being. However, Debbie was still observing differences between two naturally occurring groups. Therefore, despite her efforts to control for differences in her analysis, Debbie cannot be certain that the differences in outcomes she observed were not due to an unknown factor that she had not identified. This is why a longitudinal cohort study cannot provide evidence of a causal relationship between different events.

Experimental designs

In an **experimental** study, you take some measurements, provide some sort of treatment or intervention, then take some measurements again to assess what kind of impact the new treatment/intervention has had. **Case example C** in **Figure 11** provides an example of an experimental design.

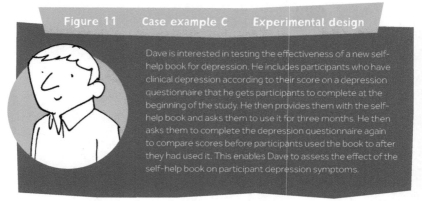

Figure 11   Case example C   Experimental design

Dave is interested in testing the effectiveness of a new self-help book for depression. He includes participants who have clinical depression according to their score on a depression questionnaire that he gets participants to complete at the beginning of the study. He then provides them with the self-help book and asks them to use it for three months. He then asks them to complete the depression questionnaire again to compare scores before participants used the book to after they had used it. This enables Dave to assess the effect of the self-help book on participant depression symptoms.

Discussion point 1: Read through **case example C** in **Figure 11** again. What do you think were the problems with the design of Dave's project that would have made it difficult to prove his self-help book had resulted in reduced depression?

Dave measured people's depression scores before and after providing them with his new depression self-help book. Dave's study could be described as an **uncontrolled pre- and post- (or pre-post) design**. Dave didn't compare the pre- and post- effects of his self-help book on depression scores with another group of people with depression who had not used his self-help book. In an experimental design, the comparison group of participants that do not receive the new treatment/therapy are known as the **control group**. If Dave had used a control group his design would be called a **controlled pre- and post- (or pre-post) design**.

Using a controlled pre- and post- (or pre-post) design,
Dave could demonstrate:

a) that there was a greater improvement in depression scores in the people who had used his self-help book than in those in the control group who hadn't used it.

b) that the average improvement in depression scores over the control group had not occurred by chance (this is calculated using statistical calculations, see Chapter 4. Generally, if researchers can demonstrate that there was less than 5% chance the difference occurred by chance, the result is accepted as **significant**).

Using this approach, Dave would have some evidence of a link between his self-help book and reduced depression. However, even if Dave could say that he had met these two requirements, he could not say that there were not important differences between those who received his book and those in the control group that might explain the differences in depression scores at the end of the study (rather than the benefit of his self-help book). Dave could have ensured that both groups were the same by randomly allocating people to either receive his book or to the control group at the beginning of the study. This process is called **randomisation** which is a process of allocating participants to the treatment or control group effectively on the basis of a coin-toss. Along with the use of a control group, randomisation is one of the key features of a **randomised controlled trial**.

The randomised controlled trial

The randomised controlled trial (RCT) is considered the 'gold-standard' design for determining whether a **cause-effect relationship** exists between a treatment and outcomes. **Figure 12** provides a more detailed description of the features of an RCT, but the key steps involved are:

1. Select participants
2. Measure **baseline variables** (e.g. using **case example C** this would be the score on the depression questionnaire, but you would also, as a minimum, collect basic demographics such as age, gender etc.)
3. Randomise (to treatment or control group)
4. Apply intervention (e.g. using **case example C** providing participants in the treatment group with the depression self-help book)
5. Measure follow-up outcomes (e.g. using **case example B** this would involve repeating the depression questionnaire with participants in both the treatment and control groups following the three months that the treatment group used the self-help book)
6. Analyse the data

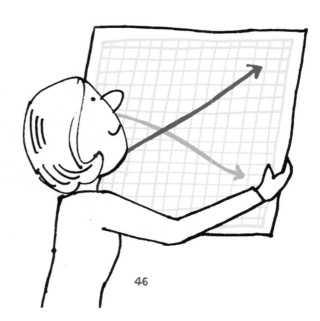

Figure 12 **Key features of RCTs**

There are two key benefits of randomisation. Firstly, it reduces bias by preventing researchers from influencing which participants receive the treatment and which are allocated to the control group. Secondly, it helps to minimise the effect of **confounding variables** (e.g. extra variables that the researchers didn't account for which can ruin an experiment by giving them incorrect results) through evening out different participant attributes between the intervention and control group. So, again, we will use Dave's self-help book as an example to make this point more clearly. If Dave was conducting a controlled study of his self-help book, a potential problem might be that he had people with more severe depression and more patients on antidepressant medication in his treatment group than in his control group. Dave could address this problem by only including people who scored as having moderate depression according to their baseline depression questionnaire and excluding anyone who was on anti-depressant treatment – but this might mean he could exclude people he was trying to help. The beauty of randomisation is that, provided you have a large enough sample (i.e. enough randomisations/coin-tosses), the laws of probability dictate that these varying attributes within your sample will gradually begin to even out, until you have two equivalent groups to compare treatment effects.

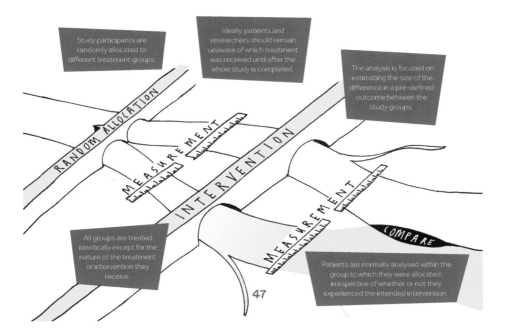

Researchers often measure the effect of a new treatment against many different outcomes, asking participants to complete many different questionnaires at baseline and at follow-up data collection points. Earlier, we discussed how researchers will accept up to 5% probability that a significant result has occurred by chance. Researchers must accept this 5% probability across all of their outcomes. Clearly then, the more outcomes researchers use, the greater the chance of getting **false-positive results**. This is why, for an RCT, researchers must select a single **primary outcome** to measure the success of their treatment/intervention against. The remaining outcomes of interest should be considered **secondary outcomes**.

The primary outcome is also used to calculate the sample size required for the trial. First, researchers must work out what would represent a **clinically significant effect** of the intervention. For example, this might be a change on a depression scale that moved a person from severe to moderate depression, or from moderate to mild depression. Once this has been agreed, researchers conduct a statistical test called a **power calculation**. This calculates the minimal sample size required to detect a significant difference between the treatment and control groups.

The final key feature of an RCT yet to be covered is the need for researchers to conduct an **intention-to-treat analysis**. This refers to the need to include all participants randomised in the final analysis of data, regardless of whether they withdrew from the trial or failed to complete questionnaires. After all, it makes sense that participants who failed to complete the study may have had less favourable experience of the intervention, so failing to include any data from these individuals provides an important source of bias in favour of the intervention. The intention-to-treat analysis works by inputting estimates of what the missing data was likely to be – this is based on earlier scores on the questionnaires that participants completed.

PPI stories from EQUIP

Next Andrew discusses his experiences of being involved in quantitative research.

Andrew's Story

One of the studies in the EQUIP research programme was an RCT of a training intervention for community mental health and social care professionals to help them improve service user and carer involvement in care planning. We compared teams trained through a new training course (our intervention) with teams who did not receive this training (our control). Service users rated different aspects of the services they received from these teams before and after training.

RCTs are a quantitative research design. Study design and data analysis were led by the research team but I played a major role in developing and delivering the new training intervention for our trial. Our team met to co-design our intervention using information gathered from a literature review (Bee et al., 2015a) and from focus groups and interviews with service users, carers and professionals (Bee et al., 2015b; Cree et al., 2015; Grundy et al.,2015).

We agreed the content and format of our training intervention and co-developed a training manual and presentation slides for a two-day training course.

I was an integral member of the team who delivered this training course to the community mental health and social care professionals participating in our trial. To prepare me for my role, I attended a 'train the trainers' course (Fraser et al., 2017), which equipped me with the practical skills for training these professionals.

I ended up delivering our new training intervention to 18 different community mental health teams. It was important to me that I use my lived experience to do this. Other service users and the carers co-facilitated group work with me. We shared positive and negative experiences of care planning, and shared ideas around good and poor practice throughout the two days.

I co-facilitated follow-up supervision with teams who were trained in our trial, and I assisted some service users to complete our outcome measure pack. Sometimes I did this over the phone and sometimes it was in person. Supporting people to complete trial outcome measures is important because it can help to make sure these people aren't excluded from health research studies. It can make these studies less daunting and increase the number of people who want to take part in a trial.

REFLECTIVE EXERCISE

- When and why might you use a randomised controlled trial?
- What are the key features of an experimental study?
- Describe an appropriate method of randomisation and describe the benefits of undertaking randomisation.

REFERENCES AND FURTHER READING

Sibbald, B. and Roland, M. (1998) Understanding controlled trials: why are randomised controlled trials important? British Medical Journal 316:201.

CHAPTER 4:
QUANTITATIVE DATA ANALYSIS

Patrick Callaghan and Penny Bee

CHAPTER OVERVIEW

Quantitative data analysis makes sense of numerical data. We often refer to quantitative data analysis as statistical analysis, and you may see this term used in published research papers. We can use numbers to summarise the experiences or characteristics of a group of participants, for example their average age or the number of symptoms they report. We can also use numbers to look at people's behaviours, experiences and views, for example the number of people using mental health services or the proportion of people who are satisfied with their care. Perhaps most importantly, we can use numbers to look at differences between groups of people or the same group over time. This can help us understand the effect of new treatment or policy initiatives, both in terms of the type of effect (e.g. does a new policy make things better, worse or leave things unchanged?) and the size of its impact (e.g. are any changes big enough to be meaningful or could they have happened just by chance?). For some people, numbers and statistics are reassuring, but for others they can be baffling. In this chapter, we will explore some of the different approaches to analysing numerical data, explore the difference between descriptive and inferential statistics, and highlight some of the ways in which you can begin to interpret research data presented as numbers.

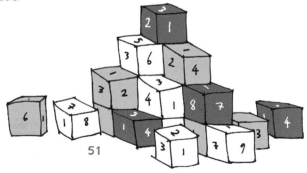

LEARNING OBJECTIVES

By the end of this chapter you should be able to:

1. Describe the difference between descriptive and inferential statistics
2. Understand key concepts in analysing quantitative (numerical) data
3. Discuss whether results are statistically and/or clinically significant

INTRODUCTION

The difference between descriptive and inferential statistics

There are two broad categories of numerical data analyses that researchers are likely to find themselves doing: descriptive and inferential statistics.

Descriptive statistics are used to organise and describe a dataset. Often, descriptive statistics are used to describe the characteristics of a group of research participants (e.g. the range, or most common, score that study participants achieved on a scale measuring anxiety symptoms), but they can also be used to describe other things, such as the characteristics of a health service. A good example here might be the average waiting time for treatment or the rate of staff turnover on a hospital ward. Descriptive statistics are important because they help us to visualise or understand what our data is showing. However, they apply only to the data we have collected and do not allow us to draw any bigger conclusions.

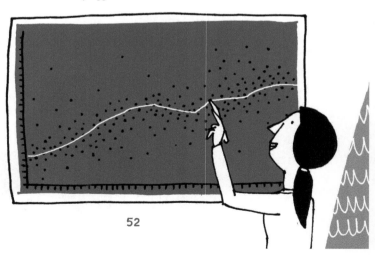

Inferential statistics are used to compare two or more datasets and to explore whether and how they differ. Inferential statistics are used to infer something. They allow us to generalise beyond our own dataset to draw conclusions about a bigger population. They can help us to understand, for example, the impact of a change in health policy or the effect of a new treatment.

Descriptive statistics

Many different statistics can be used to describe a dataset, but the main ones that you are likely to see reported are the median, mode, mean, standard deviation and range. The mean, median and mode are often referred to as measures of central tendency. They give us an indication of the central position in a dataset. Standard deviation and range are measures of spread – they tell us how much the data varies, or spreads out around this point.

Let's use **case example A** from the previous chapter to help us work through these concepts.

Case example A

Paul is interested in looking at how common clinical depression is in people living in Manchester. He sends out a questionnaire to everyone in Manchester measuring their levels of depression and collecting basic demographics such as age, gender, ethnicity, sexuality, accommodation and employment status.

The first 11 people to return their questionnaires report their age (in years) as follows:

22, 49, 33, 41, 87, 18, 33, 54, 40, 33, 72

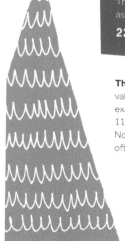

The mean is the average of all the data. We calculate this by adding up all the values in our list and then dividing by the number of values we have. In our example, the mean is (18+22+33+33+33+40+41+49+54+72+87) divided by 11 = 43.8. This means the average age for our 11 participants is 43.8 years. Note that the mean does not have to be a number from the original list, and often it isn't.

The median is the middle value of the set, when all values are placed in size order. To find the median in the example above, first write all the numbers in order (e.g. 18, 22, 33, 33, 33, 40, 41, 49, 54, 72, 87). The middle value, the median, is 40.

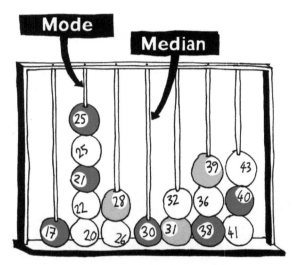

The mode is the most common value or most frequent response in a set. In our example, the mode is 33.

The standard deviation (SD) indicates how much the values in our dataset are clustered (or not clustered around the mean). The mean score on a depression symptom scale might be 7, but not everyone will score 7. Some will score higher and others will score lower. The standard deviation tells us how spread out people's scores are likely to be. A low standard deviation indicates that most data points tend to be close to the mean, while a high standard deviation indicates that the data points are spread out wider. This helps us to decide how variable people's responses are. Usually, we calculate standard deviation with the help of a calculator or a computer package, such as Excel.

The range refers to the difference between the largest and smallest value for a measure, so in the age example above, the range would be 87 - 18 = 69 years. Often research reports will just write the range out in full, e.g. 18 - 87 years.

So what do we use and when?

Although not impossible, it is more unusual to see a research report that includes all of the above statistics. Typically, a published study will report either the mean and the standard deviation (mean, SD), or the median and the range (median, range).

The choice of what to present will partially be based on the type of data we have.

There are four different types of numerical data **(Figure 13)**:

Figure 13 Levels of measurement with examples

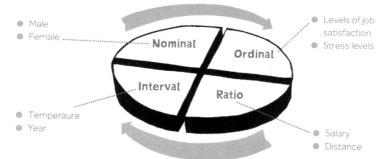

- Male
- Female

- Levels of job satisfaction
- Stress levels

- Temperaure
- Year

- Salary
- Distance

Nominal data refers to data that can be categorised into groups but which does not have an obvious order. One example of nominal data is gender. Often, we ask research participants to indicate if they are 'male', 'female', 'transgender' or 'without gender'. A good summary measure for these data is the mode, because it will tell us which one of these categories most participants identify with.

Ordinal data is data that can be categorised into groups but can also be ordered in a meaningful way. An example would be responses from a questionnaire where participants have been asked to indicate their level of agreement from 'strongly agree', 'agree', 'neither agree nor disagree', 'disagree' or 'strongly disagree'. We can order these response categories (e.g. from most positive to most negative), but we cannot place a value on them. The best measure for ordinal data is the median, because this will give us the response category which half the sample falls above and half the sample falls below. Presenting the median and range together will tell us the spread of responses as well as the middle response.

Interval data is data that can be ordered, and where the difference between two data points is quantifiable and meaningful. An example is air temperature. The difference between 50 and 60 degrees is the same as the difference between 60 and 70 degrees.

Finally, **ratio data** is the same as interval data, but it also has a meaningful zero score. A good example of ratio data is salary level; if person A earns £20,000 per year and person B earns £10,000, person A earns twice as much as person B. Other variables, like weight, are ratio variables, but temperature is not. You can have zero earnings, but zero degrees Celsius does not mean that a room has no temperature! Interval and ratio data are usually summarised by reporting the mean and standard deviation.

Inferential statistics

Inferential statistics are tests that allow researchers to draw conclusions from their data. Whilst descriptive statistics summarise the characteristics of a group of people or things that have been measured (our study sample), inferential statistics allow us to use these data to estimate characteristics for a bigger population. This means we can draw conclusions beyond the actual group of people that have been measured.

Inferential statistics are important in research because it is never possible to obtain a measurement from everyone – not everyone wants to take part in research and even if they did, the time and cost commitment would be enormous. So instead we work with a manageable group of research volunteers. They provide their data and we use this to estimate the values that would be measured in a population.

To ensure our study group is as similar as possible to the wider population, we select or sample our volunteers carefully. However, we have to accept that there will always be some degree of uncertainty in our estimates (because we cannot measure the whole population) and this means our conclusions are to some extent always going to be our 'best guess.'

How do we know when our guess is good enough?

Deciding how many people to collect data from in a research study is not easy. People (and patients) are individuals, which means that their characteristics, outcomes and responses to treatment are likely to vary. If we only had one or two people in our sample, we would probably not be confident that their data was representative of the whole population. Similarly, if we had everyone in our sample, we could not afford to run our study and we would probably be exhausted. We therefore need to strike a balance.

Research studies, and particularly randomised controlled trials, often use a special calculation called a **power calculation** to decide how many people we need to recruit to a study. A study must have sufficient power to infer the correct result. The minimum power level which is normally accepted is 80%, which means that a study would have an 8 in 10 chance of detecting a relationship or difference between groups (assuming a difference exists). If a study has less than 80% power, then these genuine differences may not be picked up. The big risk here is that a research team would conclude that there was no relationship between variables (e.g. no effect of a new treatment) when in fact there was.

Using inferential statistics to interpret research data: Understanding variables

All research studies examine some kind of characteristic or variable. In research, a variable is not only something that we can measure, but also something that we can manipulate or control for (if we want to do so).

An **independent variable**, sometimes called a predictor variable, is a variable that is being manipulated in an experiment. It is manipulated to observe the effect on a **dependent variable**, which can also be called an outcome variable.

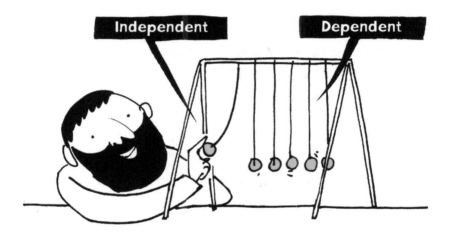

Let's return to an example used in the last chapter (Case example B).

Case example B

Dave is interested in testing the effectiveness of a new self-help book for depression. He includes participants who have clinical depression according to their score on a depression questionnaire that he gets participants to complete at the beginning of the study. He then provides them with the self-help book and asks them to use it for three months. Next he asks them to complete the depression questionnaire again to compare scores before participants used the book to after they had used it. This enables Dave to assess the effect of the self-help book on participant depression symptoms.

Discussion point 1:
Read through case example B above. What do you think were the independent and dependent variables in this study?

The aim of Dave's study is to examine whether the use of a new self-help book result in a change in depression symptoms. So, in our example, the independent variable is the use of the self-help book and the dependent variable is depressive symptoms. The dependent variable does exactly what is says – it is the variable that is dependent on the independent variable.

In experimental research, the aim is to manipulate directly the independent variable(s) and measure the effect on the dependent variable(s). Because a change in the former can be directly linked to a change in the latter, experimental research has the advantage of enabling a researcher to identify cause and effect.

In non-experimental research, the researcher does not manipulate the independent variable(s) themselves. Often this is because it is impractical or unethical to do so. For example, we might be interested in the effect of sudden trauma on people's mental health. It would be unethical to expose study participants to trauma just to study its effects. In this case we might identify a group of adults who have already experienced trauma and compare their questionnaire responses with another group who have not had this experience. Exposure to trauma is still the independent variable and mental health the dependent variable but because we have not directly manipulated the trauma variable, it is not possible to fully establish cause and effect. Instead, we focus on the strength of association or correlation between the two variables.

Using inferential statistics to interpret research data: Understanding statistical significance

Inferential statistics are used in both experimental and non-experimental designs. To make sure we interpret our data correctly, we need to consider:

- whether our results could have occurred by chance
- how meaningful our result is in the real world

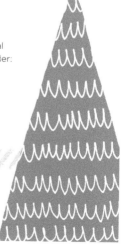

Could our results have occurred by chance?

As a general rule, researchers expect to infer one of two things from their studies: either that the independent variable has no effect on the dependent variable, or that the two are related.

The notion that there will be no effect is sometimes referred to as the **null hypothesis**. A null (or zero) hypothesis always states that there will be no relationship between the variables being compared.

In contrast, **the research hypothesis** will usually state that a relationship is expected. Researchers will use their current knowledge and previous work to predict what they think this relationship will be.

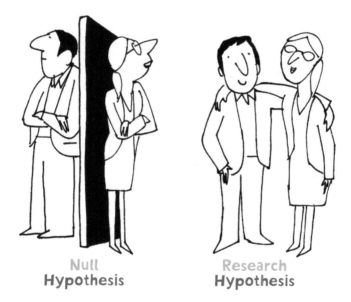

Null **Hypothesis**   Research **Hypothesis**

Research teams rely on statistical tests to analyse their data, to determine whether the null hypothesis or the research hypothesis is more likely. There are many different types of statistical tests, but all of these provide something called a **p-value**. A p-value is a measure of statistical significance. Put simply, it is the likelihood that any relationship observed between the variables is caused by something other than chance.

As the p-value gets lower (i.e. closer to zero), we are more inclined to accept a research hypothesis and to conclude that there is a relationship between our variables. A cut-off of 5% (p=0.05) or 1% (p=0.01) is conventionally used to indicate statistical significance. This means that any p-value lower than these values is normally accepted as indicating a significant and 'real' result. P-values above the cut-off (of either 0.05 or 0.01) suggest that there is unlikely to be a significant relationship between our variables, and prevent us from rejecting the null hypothesis.

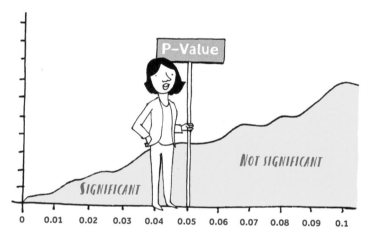

To put this into context, let's look at the EQUIP trial.
The EQUIP trial aimed to answer the following research question:

Do service users treated by a community mental health team trained in user involvement have different outcomes to those treated by teams who have not been exposed to training?

For this question, the null and research hypotheses would be set up as follows:

- **Null hypothesis:** There will be no differences in the outcomes of service users treated by trained and non-trained community mental health teams.

- **Research hypothesis:** Service users treated by a community mental health team trained in user involvement will have different outcomes to service users treated by a community mental health team that has not been exposed to training.

The research question and hypothesis suggest that service users' outcomes are somehow related to whether the community mental health team that treats them is exposed to training. Outcomes included perceived involvement in the care planning process and satisfaction with care. The EQUIP researchers collected data measuring various service user outcomes from a representative sample of service users receiving treatment from community mental health teams exposed, or otherwise, to training. When the researchers analyse these data, we would like to know if the results can be generalised to all service users, not just those who participated in the EQUIP study. This requires the use of a statistical test.

Imagine the researchers analysed the results from this trial and found that they were statistically significant at $p < 0.05$. Therefore, service users treated by a community mental health team that has been exposed to training will have different outcomes to service users treated by a community mental health team that has not been exposed to training. The null hypothesis (i.e. that there are no statistical differences) can be rejected!

A portion of the results reported by the EQUIP trial are provided below.

Outcome	Time	Usual Care			Intervention			Adjusted mean difference	P-value
		Mean	SD	N	Mean	SD	N		
HCCQ (Support for service-user autonomy)	Baseline	5.06	1.66	272	5.27	1.48	332		
	6 Months	4.93	1.78	227	5.01	1.70	269	-0.064	0.654
HADS-D (Depression)	Baseline	9.18	5.30	272	10.05	5.19	332		
	6 Months	8.90	5.81	172	9.82	5.50	208	-0.020	0.963
VSSS (Service satisfaction)	Baseline	3.58	0.62	272	3.53	0.61	332		
	6 Months	3.53	0.80	156	3.52	0.72	191	0.121	0.045

Discussion point 2:

The EQUIP trial aimed to embed shared decision-making in routine community mental health services by delivering a practical and feasible training intervention to improve service-user and carer involvement in the care planning process. They used the Health Care Climate Questionnaire (HCCQ) to measure the extent to which service-users felt supported in their care decisions. They also measured their satisfaction with community mental health services (measured by the Verona Service Satisfaction Scale, VSSS) and their levels of depression (using the Hospital Anxiety and Depression Scale, HADS-D).

Have a look at the p-values in the results table above. Do you think the results of the EQUIP trial were statistically significant? What conclusions should the team have drawn?

Answer: The results of the EQUIP trial showed that there were no statistically significant differences in HCCQ scores between the intervention and usual care groups six months after training. Depression scores were not significantly different between the two groups. The intervention group reported higher satisfaction with services compared to the control group. This difference was very small and only just statistically significant at the 5% level.

The EQUIP trial team concluded that their training intervention was not sufficient to embed shared decision-making into routine community mental health services. They concluded that successful intervention was likely to require much greater investment of resources. They suggested that the effects of a staff training intervention may take time to emerge and may only become apparent after six months.

Type 1 and type 2 errors

Sometimes, you hear research teams talk about type 1 and type 2 errors.

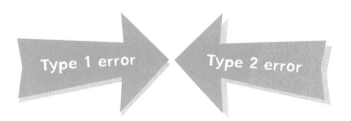

A type 1 error occurs when you conclude that two variables are related to one another when in fact they are not (a false positive). A type 2 error occurs when a researcher concludes that two variables are not related when in fact they are (a false negative).

Type 1 and type 2 errors can mislead people and might have serious consequences. Let's think for a moment about a randomised controlled trial testing the effects of a new drug. A type I error would mean that we concluded the drug was effective in treating a person's symptoms when in fact it wasn't. At the extreme, this could lead to unnecessary cost in producing and distributing the drug, unnecessary use of the drug (possibly with accompanying side-effects) and a failure to properly treat or control the target illness. A type 2 error would mean that the research team would conclude the drug was not effective, when in fact it could have helped a lot of people.

Useful ways of avoiding these errors and reducing threats to the validity of a research study are to:

- have an adequate sample size (to ensure the study has sufficient power)
- recruit a representative sample (to ensure that the sample providing data resembles the wider patient population)
- follow an appropriate research design (to minimise confounding variables)
- use reliable and valid outcome measures (to ensure that any changes are detected and recorded)

Using inferential statistics to interpret research data: Understanding clinical significance

Statistical significance is not always the same as clinical significance. Even if a result is statistically significant we still need to consider clinical significance, because this tells us how important a study's findings are likely to be in the real world.

Clinical significance, sometimes called practical significance, is our 'so what' question. To decide if a study's findings are meaningful in the real world you need to consider carefully the research questions and the measures used to collect the data. For example, if a study shows that depression scores reduce by one point after treatment, is this a useful change? The result might be statistically significant (i.e. have a p-value less than 0.05) but what does a reduction of one point really mean for patients? Could a reduction of this magnitude noticeably enhance a person's mental health, daily functioning or quality of life? Would a greater reduction in scores normally be required before people noticed an effect?

The minimal clinically important difference is the smallest difference or change in outcomes which patients perceive to be beneficial. In the context of a randomised controlled trial examining the effects of a new treatment, it is the smallest difference that would justify changing patient care.

Decisions about clinical significance are not always easy, and can require detailed knowledge of the area. However, there are some key concepts that are usually quoted in research papers that may help you in this respect:

Effect size: The effect size quantifies the difference between two or more groups. In an RCT, it is a measure of the difference in the outcomes between the experimental and control groups. Effect sizes are based on the mean and standard deviation of the outcome scores in each group, and are often standardised so that differences across several different outcome measures with different units can easily be compared. Effect sizes of 0.2 or below are usually considered small, effect sizes over 0.5 are medium and effect sizes over 0.8 are large.

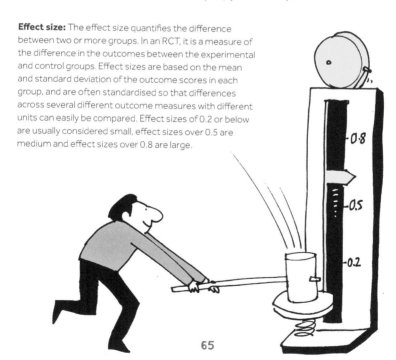

Odds ratio: An odds ratio is a bit like an effect size in that it also allows the outcomes of an experimental and control group to be compared. However, unlike an effect size, it is often used for categorical data and provides a more relative measure of effect. The odds ratio represents the odds that a particular outcome will occur given a particular context or treatment compared to the odds of the same outcome occurring in the absence of that context or treatment.

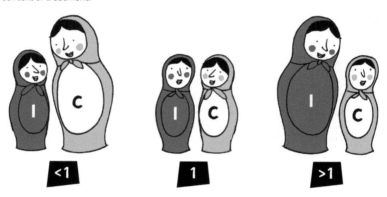

If the outcome is the same for both groups (i.e. no difference between them) then the odds ratio will be 1. The odds ratio needs to be more than 1 for the intervention to be considered better than the control. If the odds ratio is less than 1 then it means the control group is better than the intervention.

Confidence interval: Confidence intervals are used to indicate the level of uncertainty around an effect reported in a research study. Earlier we discussed how it was practically and financially impossible for a whole population to take part in a research study. Instead we recruit a 'sample' of people, our research volunteers, who provide us with data that we can use to estimate the range of values that we would be likely to see in a whole population.

Estimating means we are unlikely to measure exactly the right result. If we were to run our study lots of times, we might get a slightly different result each time. Confidence intervals help us to decide how much these different estimates are likely to vary. By having an upper and lower confidence limit, we can show the range of values between which we think the true result lies. A narrow confidence interval suggests our estimate is likely to be fairly accurate, as there is little room for it to vary. A much wider confidence interval suggests that our estimate may be less precise.

Statistical analysis plans: What are they and why do researchers have them?

Before conducting any research, a team should agree on what they would like to investigate and how they plan to go about it. Making a detailed plan before you start can help to make sure that you can answer all of your questions by the end of the study, and that you don't make any mistakes along the way.

When it comes to analysing data, we follow the same procedure: before we look at the data we make a detailed plan of everything we would look at and what statistical tests we will use to do this. It is important that this plan, sometimes called a statistical analysis plan (or SAP), is done before we look at the data. This helps to make sure that our decisions aren't influenced by what we can find.

SAPs detail all the different things that will be done to the data in order to get the results to answer our research questions. This usually includes: how we plan to summarise the data, what checks will be carried out to make sure there have been no mistakes when collecting the data, defining all the hypotheses we wish to test and exactly what statistical methods will be used to test these hypotheses. The plan also allows us to describe what might need to be done if parts of the study haven't gone to plan. For example, what will we do if some participants do not take part in the treatment or intervention they have been allocated to? What will we do if some participants do not come for their follow-up meeting and have missing data? What will we do if we look at the characteristics of the intervention and control group and find that one group is much older than the other, or has more women?

Defining all these steps ahead of time means that we can do the statistical analysis quickly at the end of the study. It also gives us a good audit trail. If someone queries what we've done, we can show them the exact steps that we planned to do and the steps that we followed. If they were to carry out the same steps, they should get the same results.

Factors influencing the choice of a statistical test

Various tests are used in research designs, the choice of which is normally made by the team's statistician. The factors influencing the choice of statistical test will depend to a large degree on the specific test; each test has assumptions that should be met before the test is used. General factors that influence choice of statistical test include:

- The research design
- Sample size and sampling method
- Number and nature of the independent and dependent variables
- The spread and pattern of people's responses to an outcome measure

Quantitative data analyses allow researchers to make sense of numerical data gathered from research. Descriptive statistics are used to organise and describe data numerically often through showing measures of central tendency (e.g. mean, median, mode) or data spread (e.g. standard deviation, range). Inferential statistics allow researchers to draw conclusions from their data by testing hypotheses using statistical tests and by calculating estimates of clinical significance such as effect sizes or odds ratios. The SAP provides a clear and transparent roadmap for data analysis, deciding which tests and levels of significance are going to be used in advance. This ensures that there is no bias in the analysis and that the statistical results of a research study can be trusted.

PPI stories from EQUIP

Next Andrew describes his experiences of being involved in quantitative data analysis.

Andrew's Story

We delivered the EQUIP training intervention to 350 professionals from 18 community mental health teams in our trial. Every time we delivered training we used a questionnaire, called the Training Acceptability Rating Scale (or TARS), to measure the acceptability and perceived impact of our work.

We collated all of the questionnaires together and analysed people' responses. We reported the percentage of people who responded and used descriptive statistics to summarise their scores. We calculated the median and range of scores for each question on the TARS.

Our results suggested that our training course was acceptable to health professionals and had had a positive impact on their attitudes, knowledge and skills. It was good to get this feedback from course attendees. I have since co-authored a paper that discusses our findings (Grundy et al, 2017).

Reflective exercise

- What is the difference between descriptive and inferential statistics and when might you use each?
- What is a power calculation and how might researchers use one?
- Define type 1 and type 2 errors.
- Consider three factors that might impact on your choice of statistical test.

REFERENCES AND FURTHER READING

Best, J. (2001) Damned lies and statistics: untangling numbers from the media, politicians and activists. Berkeley: University of California Press.

Best, J. (2004) More damned lies and statistics: how numbers confuse public issues. Berkeley: University of California Press.

Dancey, C.P., Reidy, J. and Rowe, R. (2012) Statistics for the health sciences: a non-mathematical introduction. London: Sage.

Field, A.P. (2013) Discovering statistics using IBM SPSS statistics: and sex and drugs and rock 'n' roll. London: Sage.

Grundy, A.C., Bee, P., Meade, O., Callaghan, P., Beatty, S., Ollevent, N. and Lovell, K. (2016) Bringing meaning to user involvement in mental health care planning: a qualitative exploration of service user perspectives. Journal of Psychiatric and Mental Health Nursing 23:12-21.

Haslam, S. A. and McGarty, C. (2014) Research methods and statistics in psychology (2nd edition.). London: Sage.

Huff, D. (1954) How to lie with statistics. New York: Norton & Company.

Kanji, G.K. (2006) 100 statistical tests. London: Sage.

Levitt, D.S. and Dubner, S.J. (2005) Freakonomics: a rogue economist explores the hidden side of everything. New York: William Morrow and Company.

Salsburg, D. (2001) The lady tasting tea: how statistics revolutionized science in the twentieth century. New York: W.H. Freeman.

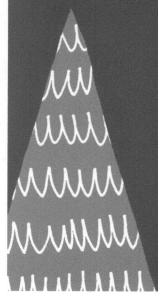

Chapter 5:
Health Economics

Linda Davies and Gemma Shields

Chapter Overview

Evidence is needed to inform and guide the choices that healthcare organisations make in relation to how budgets are spent. The associated costs and benefits of health treatments are key components of such decisions. An economic evaluation is a way of systematically identifying the costs and benefits of different health activities and comparing these to make an informed decision about the best course of action based on the evidence available. Economic evaluations can also be used to identify uncertainty around the likely costs of a particular health activity and to compare this against a 'willingness to pay' threshold, in order to judge their value for money. Economic evaluations can be done as part of randomised controlled trials or can draw on evidence taken from other sources (e.g. surveys). Similar to clinical evidence, economic evidence needs to be updated and researched as new questions arise or more evidence becomes available.

Learning Objectives

By the end of this chapter you should be able to:

1. Understand why economic evaluations are needed

2. Understand the key parts of economic evaluations and the data that feed into them

3. Be able to begin to interpret and understand the results of economic evaluations

INTRODUCTION

The cost of providing healthcare is rising. As the population grows and becomes older, this increases the demand for healthcare interventions (e.g. drugs, therapies and services). Health services have limited budgets to meet these demands and need fair and objective ways of deciding which treatments, out of all those available, offer the best value for money. Economic evaluations compare the costs and benefits of a healthcare approach to assist decision makers in making these decisions.

The results of economic evaluations help decision makers make choices between the many healthcare interventions available which in turn helps to ensure that resources are used in the most efficient way to maximise the health of a particular patient group or population. Economic evaluations bring together data on healthcare benefits and costs, as well as identifying any uncertainty that might exist in the research data. In this chapter we will explore the role of economic evaluation in healthcare, the different types of evaluations that might be conducted and how economic data can be analysed.

Making choices in healthcare

First, let's consider how we make choices in our everyday lives - we consider a number of different things, including the likely benefits and drawbacks of our choice, its associated costs and risks, and who it might involve. We also consider what we do not know but might need to know, and, if necessary, seek further information.

For example, consider buying a car. If you have more than one option, you might compare cars in terms of their costs (and whether or not these are affordable), their mileage, makes, models and from whom or where you are buying them. You will consider your options, weigh up these different pieces of information and decide which one, if any, is best for you overall. We make many decisions every day in the same way (e.g. which brand should I buy in the supermarket? How should I travel to work?).

Choices about healthcare differ from these more 'everyday' choices in a number of ways:

1. The individual receiving the care doesn't always pay (In the UK for example, the individual receiving the care doesn't always pay in full).

2. Choices about treatment availability are often made before a treatment is needed. National organisations will look at a range of evidence (e.g. evidence of treatment effectiveness, side effects and costs), to judge which treatments should be made available and recommended for use.

3. Treatment choice is influenced not only by the person who will receive the treatment, but also by their healthcare providers (e.g. your doctor will likely prescribe a treatment for you, based on the evidence and their knowledge).

The role of an economic evaluation

Generally speaking a treatment that is shown to be beneficial to service users (e.g. reducing symptoms or improving quality of life) whilst at the same time having low costs for the organisation (e.g. financial and time) would be considered the best approach.

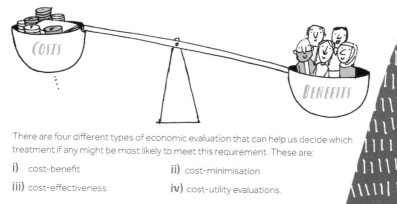

There are four different types of economic evaluation that can help us decide which treatment if any might be most likely to meet this requirement. These are:

i) cost-benefit
ii) cost-minimisation
iii) cost-effectiveness
iv) cost-utility evaluations.

In this chapter, we will focus on cost-effectiveness and cost-utility evaluations, as these are the most recommended methods for economic evaluations in England (NICE method guide, 2013).

Cost-effectiveness evaluation: This type of analysis looks at differences in the costs and health benefits of two or more interventions. It summarises this difference by producing something called 'a cost per health benefit'. Health benefits are usually measured in natural units that are relevant to an individual, e.g. life-years or the number of symptom free days obtained with treatment. A cost-effectiveness evaluation usually compares one (newer) treatment with another (older or more routine treatment), to see how much cost would need to be spent by the health service to extend life by one-year. A more cost-effective treatment would allow patients to gain the same health benefit at less cost. Cost-effectiveness analysis provide very useful information but can only explore the cost of achieving one type of health benefit at a time. This means that any other benefits obtained from the treatment, or the value placed on that benefit by an individual, may not be taken into account.

Cost-utility evaluation: A cost-utility evaluation is a type of cost-effectiveness analysis but is different because it is able to take into account more than one health benefit (e.g. symptom free days). Instead, the health benefit measure used, referred to as 'utility', takes a wider approach, often also incorporating quality of life. This is important because although a cost-effectiveness study can tell you that a treatment may prolong a person's life by one-year, it does not allow you to consider the quality of that life year for the individual. Cost-utility analyses allow you to look at both the quantity **and** quality of any health benefits.

In the UK, the National Institute for Health and Care Excellence (NICE) evaluates new health interventions. NICE states that the health effects of these new treatments should always be expressed as quality-adjusted life-years (QALYs). A QALY is a measure of both quantity of life gained, and the quality of the health that is achieved. One QALY is equal to one year of life in perfect health. QALYs are calculated by estimating the years of life available to a patient following a particular treatment, and adjusting that to take into account the quality of those years of life for an individual patient. Researchers do this by asking people about how well they undertake daily activities such as washing and dressing, how well they can get around and the pain they experience from mental or physical health symptoms.

Defining healthcare costs

Economic evaluations bring together data on healthcare benefits and costs. The type of costs required to undertake an economic evaluation in healthcare include:

- **Direct costs** - the resources needed to treat an episode of illness, to monitor changes in health or to prevent health problems for example time and salary cost of a community mental health nurse.

- **Indirect costs** - the wide range of other costs that may be affected by a person's health, for example the costs that result from not being able to work while unwell (e.g. lost wages).

Example

There are two groups of antipsychotics that are used to treat schizophrenia; First-Generation Antipsychotics (FGA) and Second-Generation Antipsychotics (SGA). First Generation Antipsychotics are an older type of antipsychotic medication than Second Generation Antipsychotics. In 1999, a trial began to compare the use of FGA versus SGA. This trial was called the CUTLASS trial.

At the time the CUTLASS trial began, doctors could only prescribe first generation drugs. However local psychiatrists and service users wanted the newer second generation drugs to be made available, as there was some evidence that they were safe to use and were effective in reducing the symptoms of schizophrenia, with less troublesome side-effects.

However there was a big cost difference; the newer SGA drugs cost the NHS around £1500-£2000 per person per year of treatment, compared to around £100 a year for the FGA.

The research trial aimed to answer the following questions:

1. Should health services have the option to provide SGA?

2. Should SGA replace any or all of the FGA?

To do this, the trial looked for any differences between FGA and SGA in relation to:

- their impact on people's health (e.g. physical and mental health symptoms)
- their side effects
- the costs of using the medication
- their cost-effectiveness
- the certainty that the results were correct.

Only direct costs were included. This covered the costs of providing the two types of antipsychotic medication and the costs of any other health services that were received (e.g. inpatient, outpatient and community care). The benefits of the FGA and SGA were summarised as QALYs, calculated by looking at differences in survival and quality of life.

Conducting an economic evaluation

Figure 14 provides a visual summary of a cost-effectiveness or cost-utility economic evaluation. The first box shows that decision makers are often faced with a choice. In our example, the choice was between two groups of drugs for schizophrenia (FGA vs SGA).

By comparing the costs of both options and their health benefits, the economic evaluation produces an important result. This result is called an Incremental Cost-Effectiveness Ratio (ICER). Although this sounds very technical, it is simply the difference in the costs of the two treatments divided by the difference in benefit. The ICER gives you the cost per health benefit. What the health benefit is depends very much on whether you are doing a cost-effectiveness analysis or, e.g. you could have calculated the cost of each symptom free day gained or a cost-utility nalysis e.g. the cost of a quality adjusted life year (QALY).

Figure 14 Diagram of cost-effectiveness

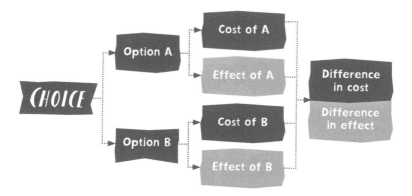

Willingness to pay

Often the cost of the health benefits identified through an economic analysis will be compared with the maximum amount that a healthcare organisation is prepared to pay to improve the health of its community (known as the willingness to pay threshold - WTPT). This helps to determine if a new intervention offers any demonstrable value over an old one, and whether it should be used. Willingness to pay thresholds are often decided in advance by organisations and reviewed annually. For example, NICE guidelines refer to a threshold of £20,000-£30,000 per QALY. Therefore, hypothetically speaking, if a new drug treatment was found to cost the NHS £11,000 a year in order for a service user to gain one additional QALY then it would be considered cost effective in line with NICE guidelines. It could then be considered a viable option for policy makers and service providers.

An example of this type of analysis is shown in Figure 15. If a new intervention is found to be both cheaper and more effective, it is said to 'dominate' the comparator (usual care) and lies in the bottom right-hand quarter. If it is more expensive and less effective it will instead be' dominated by' the comparator and fall in the top left quadrant. ICERs in the remaining quadrants need to be judged against the pre-determined willingness to pay threshold. ICERs falling under the line will be judged to be cost effective, whereas ICERs above the line will not.

Figure 15 Cost-effectiveness plane

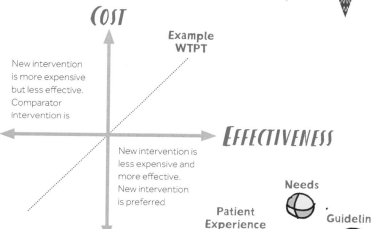

Key: WTPT, willingness to pay threshold.

In the CUTLASS trial, FGAs were found to cost slightly less than SGA (£18,858 vs. £20,118). They were also found to marginally increase QALYs (0.74 vs. 0.67). This means that, on paper at least, the first generation drugs look like the best choice for health services to provide. However, because the differences in costs and benefits were relatively small, another possible option might be to offer both types of drug and let service users decide for themselves which medication they prefer. It is important to remember the ICER result is at best an estimate, and a level of uncertainty will always be present in the data. For this reason the ICER is an important, but not the only, consideration in commissioning health services and many other factors, such as patient need and experience can influence decision making.

EXAMPLE

TABLE 3 CUTLASS TRIAL ECONOMIC EVALUATION RESULTS

One year costs and QALYs	FGA	SGA	Net value (FGA minus SGA)
Cost of all services	£18,858	£20,118	-£116
QALYs	0.74	0.67	0.04

Acknowledging uncertainty in economic evaluations

Estimating costs and health benefits is not an exact science and therefore there is always some uncertainty around the data informing an economic evaluation. Estimates of costs and benefits may differ across populations and healthcare providers. Plotting the cost-utility results produced with different estimates of costs and effects can help to judge whether or not an intervention is, on balance, likely to be cost-effective. In a similar vein, it is important that economic evaluations are critically evaluated to ensure that they are robust and relevant to decision makers. Like trials, economic evaluations can be prone to bias and this can raise the level of uncertainty associated with their findings.

It is very important that economic evidence, like any other research evidence, is updated regularly as and when decision and policy makers have new questions, and as healthcare systems, population needs and treatment options evolve. The research cycle (shown in Figure 16) describes how economic data, like other forms of data can contribute to the flow of knowledge and new research aims.

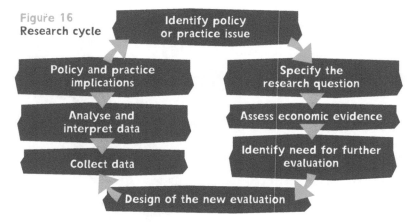

Figure 16
Research cycle

PPI stories from EQUIP

Next, Andrew describes his experiences on being involved in the health economic elements of the EQUIP project.

Andrew's Story

The EQUIP health economists devised a survey to try to capture service user's and carer's preferences for involvement in care planning. The economists needed to collect a lot of detailed information to do this, which meant the survey was not always straightforward to understand or complete.

It was incredibly important that I and my colleagues were involved in this aspect of the EQUIP study as we were able to contribute a lot.

We produced a lay-summary of the study's aims, and we piloted the survey, which gave us a good idea of how long it would take to complete, and what issues might be encountered in its completion. We revised the survey where we could to make it as user-friendly as possible. We attended outpatient clinics to give out the study packs and assist people in completing the survey where needed. We also utilised our existing networks, including social media and our contacts with local service user and carer groups, to publicise the study. Together, these activities help the study team to collect all the data that they needed.

REFLECTIVE EXERCISE

Think of a healthcare technology and imagine that you are designing an economic evaluation for that technology.

- What would it be compared to?
- What costs would you include in the economic evaluation?
- What health benefits would you include in the economic evaluation?

REFERENCES AND FURTHER READING

For interested readers who want to learn about economic evaluation in more detail, we recommend:

Drummond, Michael F., Sculpher, Mark J., Claxton, Karl, Stoddart, Greg L., and Torrance George W (2015). Methods for the economic evaluation of health care programmes. Oxford University Press.

CHAPTER 6:
PSYCHOMETRICS: DESIGNING AND ROAD TESTING NEW MEASUREMENT SCALES

Professor Patrick Callaghan

CHAPTER OVERVIEW

Measurement scales (questionnaires) are often used in quantitative research to summarise the experiences of a group of participants, for example the number and range of symptoms they report, or their level of satisfaction with their care. We can use these questionnaires once to get a snap-shot of people's scores at one point in time, or we can ask people to complete them on more than one occasion to see if their scores change. For example, if we compare people's symptom scores before and after treatment we can get an idea of whether or not the treatment they were given helped. To do this, it is essential that the scale or questionnaire that we are using has been designed to measure the outcome that we want to measure, and that it has been road-tested on a similar group of people to make sure they understand it, like it and can complete it in a way that works.

Most existing scales have been designed by clinicians, academics and researchers and often focus on people's 'clinical' outcomes, e.g. their symptoms. These clinical measures are often criticised by service users, especially if they do not tap into the priorities of the service users themselves (Crawford, 2011). As a result, we have seen increasing emphasis placed on the development of Patient Reported Outcomes Measures (PROMS). These tend to be less focussed on symptoms and more on the everyday experiences of people using services. They are much more likely to be designed and developed in collaboration with service users. The EQUIP research project developed a good quality PROM for assessing user and carer involvement in care planning, the first such measure of its kind in mental health.

This chapter will examine the origins of measurement scales in research by considering the science of psychological testing. In particular the chapter will provide a brief definition of a measurement scale, outline why scales are used, examine the design and evaluation of scales, discuss what the responses to scales mean, outline advantages and limitations of their use, and provide examples of measurement scales developed and used in the EQUIP project and other published mental health research.

LEARNING OBJECTIVES

By the end of this chapter you should be able to:

1. Understand more about the origins of measurement scales in research
2. Understand what might influence our choice of measurement scales
3. Begin to understand how new measurement scales are designed, developed and evaluated.

INTRODUCTION

Psychometrics is the term used to describe the science of psychological testing and is concerned with the measurement of mental and behavioural processes. Objective measurement is at the heart of psychometrics. In quantitative research this commonly involves the use of measurement scales, often referred to as questionnaires. The use of measurement scales is widespread in quantitative research. However, prior to their use, researchers must ensure that such scales are robust, i.e. they are reliable and valid. Using scales that are unreliable or invalid is a major threat to the integrity of quantitative research.

Measurement scales

A measurement scale is a device for measuring pre-specified outcomes e.g. a person's reported state of mind, behaviour, performance, attitudes, intentions, abilities, personality, beliefs, cognitive functioning or style, preferences or coping style. The term measurement scale is often used interchangeably with rating scale, test, inventory, questionnaire or measure. Measurement scales can be used to look at relationships between different characteristics. For example, in a recent study exploring the relationship between resilience and depression, the researchers used a specific measurement scale – The Connor-Davidson-Resilience Scale [CD- RISC] - to quantify each participant's resilience (Smith, 2009). They can also be used to provide an assessment of people as a baseline against which to measure the success of interventions, e.g. the use of the Positive and Negative Symptoms Scale (PANSS) to assess the effect of Cognitive Behaviour Therapy on psychotic symptoms. Scales may also be used to measure the behaviour of others, for example rating the severity of observed aggression using the Staff Observation Aggression Scale – Revised (SOAS-R), or as an assessment of performance, e.g. using an appraisal tool.

Clinical vs patient-reported measures

Outcome measures can have clinical utility and meaning (e.g. classification of illness severity), or they can be used to understand more about patient experience and the treatment outcomes that are most important to patients. Patient-reported outcomes (PROMs) and patient reported experience measures (PREMs) collect information directly from patients, without interpretation by health professionals.

This means that they often reflect the priorities and treatment outcomes that are most relevant to patients themselves. PROMs can include measures of symptoms, daily activities, and functioning. They can also be used to measure patient satisfaction or treatment preferences.

Modern healthcare systems place considerable emphasis on delivering high quality, patient-centred care, and as such PROMs have enormous potential to trigger changes in healthcare delivery. Developing meaningful and useful PROMs and adopting these tools in research studies represents a major step forward in assessing the outcomes of new interventions and approaches to care.

The evaluation of measurement scales

Measurement scales are critically important to data collection and interpretation, and must accurately measure what they propose to measure, i.e. they must have sound psychometric properties. The psychometric properties of a scale are usually established during its design and development phases.

Using scales whose reliability and validity are weak is a major threat to the validity of research, is unethical, and may produce false results. Therefore, measurement scales need to be reliable and valid. There are some key issues to consider when assessing measurement scales for use in a research study.

1. Minimising Error

All measurement scales are subject to error of some degree or another. Two types of errors are possible: systematic and random. Systematic error is less of a problem because it is predictable and affects all scales, like the error associated with a watch that loses a few seconds every 24 hours. Random error, however, is a major problem as it occurs randomly and does not happen with every participant who uses the scale. Sources of random error may include participants being upset, tired, or ill when completing the measure, or because questions are difficult to understand or answer, and instructions on how to use the measure are poor. The way in which different questions relate to each other and influence how a particular person might answer them can also cause random error.

For any measure a person has a true score; the score that a person obtains on a scale on any occasion is an obtained score (a close estimate). Testing participants on several occasions to get more than one obtained score and taking their average score gives us the best estimate and moves us closer to their true score, i.e. it reduces random error.

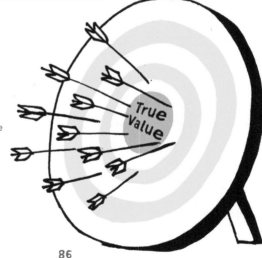

2. Establishing cut-off scores

A cut-off score is the minimum score used to decide a person's status. For example, the Beck Depression Inventory has three cut-off scores, used to indicate mild, moderate or severe depression levels. The Angoff method (Goodwin, 1996) is the most widely used method in determining cut-off scores and involves asking experts to rate the probability that the item will be an accurate measure of the concept. If the wrong cut-off score is used, a person's outcome status may not be correctly recorded.

3. Ensuring reliability

Reliability is the degree of stability with which a scale measures what it is designed to measure. There are several different types of reliability that you might come across:

- internal consistency
- test-retest
- parallel form
- inter-rater

Internal consistency reliability refers to how well individual items (or questions) in a scale measure the same thing. Internal consistency is determined by calculating a Cronbach's Alpha score between 0 and 1. A score closer to one indicates that the scale has a high internal consistency; a minimally acceptable internal consistency is around 0.75.

Test-retest reliability refers to the likelihood that a scale will measure the same score in the same participant, on more than one occasion, assuming that their circumstances have not changed. It is measured on a scale from 0 (low reliability) to 1 (high reliability) and involves linking the sets of scores of participants taking the same test at two different times. For test-retest reliability to be meaningful, it should have been tested in a representative sample of least a 100 participants, 3 months apart (Kline, 2000a).

Parallel form reliability is tested using different versions of the same measure. Participants' scores are linked and compared between the different versions to check if their responses are stable.

Inter-rater reliability is the relationship between ratings of two independent raters of the same behaviour. To assess the reliability of scales we use scores from 0 = low reliability to 1 = the highest level of reliability.

4. Ensuring Validity

Validity is the degree to which a scale measures what it was designed to measure. Different types of validity exist and include face, concurrent, content, construct, incremental and predictive (criterion) validity. As with reliability, large, representative samples are needed to establish validity.

Face validity is not really a form of validity as it only measures the participants' perception of what a scale is measuring. Nevertheless, it is important because if a participant questions the face validity of a scale they may not cooperate and this will jeopardise attempts to establish 'true' validity. Face validity can be established by simply asking participants what they think the scale is measuring.

Concurrent validity is assessed by linking scores on one scale with another measure of the same construct. This is problematic because it requires a benchmark measure of the construct (e.g. finding a scale you know measures the construct accurately) and these are extremely difficult to find.

Of course, if a benchmark measure exists it would call into question the need for a new measure. For example, in mental health we can establish concurrent validity by asking people to complete the Beck Depression Inventory and another depression measure and see how the scores to both compare.

Predictive validity occurs when people's scores on one scale predict other related things. For example, because we know anxiety and depression are often strongly related, we could assess whether high scores on the Beck Depression Inventory (BDI) will predict the likelihood that a person will also be anxious.

Construct validity is the extent to which items on a measurement scale are accurate indicators of the variable being measured. To understand construct validity a bit more, it is helpful to know what is meant by constructs. Constructs are abstract attributes that do not exist in a physical sense, such as intelligence, happiness, motivation, attitudes, self-esteem, but which are common areas of assessment for researchers and clinicians. Thus, reliable and valid measures of these constructs are necessary. The construct is the theory that underlies the measure. For example, The Eysenck Personality Questionnaire has construct validity because it is based on Eysenck's personality theory.

Content validity is the extent to which a measure evaluates all the different aspects of a construct. For example a depression scale would ideally measure a person's feelings, thoughts and behaviours to judge whether or not a person is living with depression, and/is how severe this depression is. Content Validity can be established by asking a panel of experts in the field what they think should be included. McDowell (2006) suggests that any effort to improve content validity must consider how well items are presented to participants and how well responses are recorded as well as the content of the questions themselves.

Factor Analysis (FA) a type of statistical procedure, can also be useful for assessing content validity as it allows you to determine whether items on a scale are measuring the same underlying construct. Items that measure the same thing 'group together' in the test. It is these groups, also known as 'factors', which give the test its name. FA has two main types: Exploratory and confirmatory. Exploratory Factor Analysis (EFA) is used when researchers develop a new scale and wish to test the number and meaning of factors that the items on the new scale measure. Confirmatory Factor Analysis (CFA) which looks at whether the factors identified for an existing scale in one population, also remain valid when the scale is tested on a different population.

Interpreting people's responses to measurement scales

Measurement scales generate a score that is obtained by assessing a person's response, or an assessor's response, to the items on the scale. Responses are often anchored by a numerical value and this value often indicates the strength of the response. For example, a pain score might measure the severity of a person's pain on a scale from 0 (none) to 10 (unbearable), or an opinion-based score might measure a person's level of agreement from 0 (strongly disagree) to 5 (strongly agree). Respondents' scores might also be used to categorise their responses, e.g. responses to the Beck Depression Inventory (BDI) generate a score that categorises depression as mild, moderate or severe.

Responses to measurement scales are seldom used as the sole criterion in diagnosis; they may be a useful adjunct to other forms of assessment, such as a diagnostic interview. For example, responses to the Eating Disorder Inventory (EDI) do not indicate that the respondent has an Eating Disorder; the responses indicate any similarity in scores between the respondent and the responses of those who have been diagnosed with an Eating Disorder. Because of the dangers in misinterpreting scores, many scales and tests are authorised for use only by those with a minimum level of training.

The advantages and limitations of using a measurement scale

Most measurement scales are relatively easy to use and may be a useful adjunct to other forms of assessment in helping clinicians arrive at accurate diagnoses. Scores generated from measurement scales provide a baseline, or benchmark against which the efficacy and effectiveness of interventions may be assessed. Measurement scales may be useful in tracking how much peoples' symptoms change over time or with treatment, or to provide feedback to people on the level of performance they attain on a particular task.

Responses to measurement scales are usually restricted to answering the items on the scale; therefore they may not reflect the totality of the respondents' views. Often measurement scales require that the user is trained in their use and this can be costly and time consuming. Also, it is difficult to interpret what respondents' scores on scales actually mean in practice; if a person's symptom score decreases by 1, is this meaningful? How much might a numerical score need to change before the person themselves feels they are better? It is also not easy to arrive at meaningful cut-off scores in order to categorise responses. For example, on the BDI, the categorization of someone as mildly, moderately or severely depressed hinges on small differences in scores. Depression in reality is seldom so easy to categorise.

Evaluating and Quantifying User and Carer Involvement in Mental Health Care Planning (EQUIP): Co-development of a new Patient-Reported Outcome Measure (PROM)

Items for the EQUIP PROM were developed from 74 interviews and 9 focus groups conducted with service users, carers and mental health professionals recruited from two large NHS Trusts. From these data, 70 items (potentially relevant questions) were developed.

First, face validity was examined with a mixed sample of 16 members of a Service User and Carer Advisory Group (SUCAG). Nine items were rejected by this panel as not being useful or relevant to what was being measured. The remaining 61 items comprised the first draft scale.

Members of the SUCAG were asked to comment on potential response formats for the scale. Consensus was reached for a 5-point Likert scale, with named anchors of 'Strongly disagree' at one end and 'Strongly agree' at the other. A middle neutral value with the category label "Neither agree nor disagree" was included (Figure 17 the EQUIP PROM).

Figure 17

	Strongly agree	Agree	Neither agree nor disagree	Disagree	Strongly disagree
The care plan has a clear objective					
I am satisfied with the care plan					
I am happy with all of the information on the care plan					
The contents of the care plan were agreed on					
Care is received as it is described in the care plan					
The care plan is helpful					
My preferences for care are included in the care plan					
The care plan is personalised					
The care plan addresses important issues					
The care plan helps me to manage risk					
The information provided in the care plan is complete					
The care plan is worded in a respectful way					
Important decisions are explained to me					
The care plan caters for all the important aspects of my life					

Following initial development with service users and carers, the 61-item scale was piloted with a sample of 402 service users and carers. The completed questionnaires were anonymised and the data were entered into a computer package. The team applied different forms of psychometric testing to identify unnecessary questions, remove questions that were difficult for people to answer and reduce the length of the scale. This analysis showed that there were 14 important items that needed to be retained in the final scale.

Additional psychometric testing confirmed that the final 14-item PROM was scientifically reliable and valid, and that it could be used by any mental health service seeking to assess the involvement of service users and carers in care planning.

Conclusion

Psychometrics allow researchers to illustrate how mental health care and outcomes associated with it can be measured. In order to measure mental health care and its outcomes using the principles and science of psychometrics, a measure must be a reliable and valid. The reliability and validity of measurement scales is an essential requirement to establish the accuracy of scales and reduce threats to research integrity. Psychometric testing was used in the EQUIP research project to develop a co-produced, robust Patient-Related Outcome Measure for assessing user and carer involvement in care planning, the first such measure of its kind in mental health.

PPI stories from EQUIP

Next, Joe and Andrew share with you their experiences of being involved in the EQUIP PROM.

JOE'S STORY

From the outset of EQUIP the service users and carers who were involved in the programme felt that there were no adequate measures for assessing involvement in mental health care planning.

As a group we reviewed lots of existing measures. We had some intense conversations about the wording of questionnaires, and these discussions highlighted how loaded some words can be.

The term 'Service-users' is a good example. Are we users? Sometimes it can sound pejorative, as in he's just a user, a taker never a giver. The word 'relapse' is another. For some people it fittingly described what can happen but for others it felt like a rebuke. Getting the language of the questionnaire right was so important, and like a lot of things in life there were no simple right/wrong answers.

As a group we felt strongly that we needed to devise a new measure, one that was both relevant and acceptable to service users. Our opinions fed straight into the design of the EQUIP programme - the team included a whole study dedicated to developing and testing a new questionnaire to measure service user involvement in care planning.

ANDREW'S STORY

We knew that developing a new patient-reported outcome measure for service involvement in care planning was important.

To begin with, I co-facilitated focus groups and interviews with service users, carers, and mental health professionals. With people's consent, we recorded these discussions and transcribed what everybody said. By reviewing the written transcripts, we were able to identify the key components and priorities for service-user involvement in care planning.

We then devised a new Patient Reported Outcome Measure (or PROM) – a questionnaire, completed by a service user, to measure quality involvement in care planning. We met as a group to draft the questions and discuss how they should be worded.

As PPI representatives and researchers, we used our existing networks, including social media and our contacts with local service user and carer groups, to get as many people as possible to complete our new questionnaire. This gave us lots of data and meant that we could validate the measure properly.

We have been able to develop a short 14-item questionnaire that is valid and reliable, and can be used by researchers and health services to measure service-user involvement in care planning.

I thoroughly enjoyed this experience, and am proud of our measure. I've since gone onto to assist in the design of other PROMs for mental health service users.

REFLECTIVE EXERCISE

- Give examples of three different types of measurement scales that might be used in mental health.
- Define, in your own words the terms reliability and validity.
- Why is it important to ensure reliability and validity of measurement scales?

ALLIED EQUIP PAPERS

Bee, P., Gibbons C., Callaghan, P., Fraser, C. and Lovell, K. (2016) Evaluating and quantifying user and carer involvement in mental health care planning (EQUIP): Co-development of a new patient-reported outcome measure. PLOS One 3:e0149973. Doi:10.1371/journal.pone.0149973.

REFERENCES AND FURTHER READING

Anastasi, A. (1988) Psychological Testing. New York: MacMillan, 6th Ed.

Andrews, G. and Jenkins, R. (1999) Management of Mental Disorders. Sydney: WHO Collaborating Centre for Mental Health and Substance Misuse, UK.

Bee, P., Gibbons C., Callaghan, P., Fraser, C. and Lovell, K. (2016) Evaluating and quantifying user and carer involvement in mental health care planning (EQUIP): Co-development of a new patient-reported outcome measure.

Goodwin, L. D. (1996) Determining Cut-Off Scores. Research in Nursing and Health 19: 249-256.

Kline, P. (1994) An Easy Guide to Factor Analysis. London: Routledge.

Kline, P. (2000a) The Handbook of Psychological Testing. London: Routledge, 2nd Ed.

Kline, P. (2000b) The New Psychometrics: Science, Psychology and Measurement London: Routledge.

Murphy, K. R. and Davidshofer, C.O. (1994) Psychological Testing: Principles and Applications. Engelwood Cliffs, NJ: Prentice-Hall, 3rd Ed.

PLOS One 3:e0149973. Doi:10.1371/journal.pone.0149973.

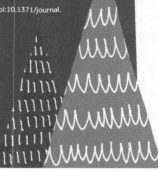

Chapter 7:
Introduction to Qualitative Research Methods

Helen Brooks, Penny Bee and Anne Rogers

Chapter Overview

The term 'qualitative research' encompasses a wide range of different methods. What underpins these is a shared aim of understanding the meaning people attribute to experiences in their lives. It has been defined as an 'interpretive approach concerned with understanding the meanings which people attach to actions, decisions, beliefs, values within their social world' (Ritchie and Lewis, 2003). Three main types of qualitative research methods were used within the EQUIP programme of work and these will form the focus of the current chapter: in-depth interviews, focus groups and observations. Throughout the chapter, the authors will refer to allied publications resulting from EQUIP as a way of providing examples of real life research to support the description of the methodological approaches provided.

Learning Objectives

By the end of this chapter you should be able to:

1. Understand different types of qualitative research methods

2. Understand the factors that influence the choice of appropriate qualitative research method

3. Understand how to carry out research utilising qualitative research methods

INTRODUCTION

What is qualitative research and why and when should we use it?

Qualitative research provides an understanding of a topic in its contextual setting giving explanations and accounts of why people do the things they do. It can also help evaluate the effectiveness of interventions and aid the development of theories and strategies. Qualitative research can be:

- undertaken independently in its own right
- as part of a bigger study or trial, to provide deeper understanding of the quantitative (numerical) results
- be used to support the development of quantitative studies by informing or testing survey content and to explore the implementation of quantitative studies

In relation to EQUIP, qualitative research methods were used to explore current service user and carer involvement in mental health care planning from the perspectives of service users (Grundy et al., 2016), carers (Cree et al., 2015), professionals (Bee et al., 2015) and other relevant stakeholders (Brooks et al., 2015) to inform the design of an intervention to enhance involvement in care planning. They were also used to inform the design of the randomised controlled trial and to examine the barriers and facilitators to the implementation and embedding of the user/carer involved care planning training intervention within the 10 mental health sites included in the trial (Figure 18).

Figure 18 Examples of EQUIP research questions addressed using qualitative research methods

- How do service users conceptualise care planning involvement?
- How can meaningful service user and carer involvement be instilled in the care planning process?
- What is the role and influence of individuals, teams and organisational factors in achieving high quality user-involved care planning?
- What are carers' experiences of the care planning process for people with severe mental illness?
- What are professionals' perceptions and experiences of delivering mental health care planning and involving service users and carers in decisions about their care?
- What factors might promote or inhibit the routine incorporation of user- and carer-led planning within mental health services?

Quantitative versus qualitative research

Quantitative methods relate to numbers. Data is collected in numerical form and presented in terms of descriptive or inferential statistics (see Chapter 4). In qualitative research, the focus is different. Data is often words, text and pictures and focuses on values, processes, experiences, language and meaning.

Quantitative methods of design and analysis are set before the study begins with data collected and analysed against these pre-set parameters. Qualitative methodology is a more iterative process, which is developed and refined during the course of a study. Positivist strategies alone are unlikely to fully address the complex process of service-user and carer involvement in mental health care planning.

Establishing the complex reality of mental health

Qualitative methods are particularly suited to the study of mental health given their ability to explore personal perspectives on illness, and individual experiences of health services and treatments. We know that the experience of mental illness and recovery is a complicated and personal one, but also that wider factors affect it, such as social networks, past experience and the environment in which people live (Brooks et al., 2014). Qualitative techniques can help us to understand this complexity by providing an holistic account and critical understanding of individuals' views and actions within the social world they inhabit.

Service-user and carer involvement in qualitative research

There are a range of ways in which service users and carers can and should be involved in undertaking qualitative research. Involvement enhances the quality of the research undertaken including the data collected and the analysis undertaken (see chapter 8: Introduction to qualitative data analysis). In EQUIP researchers worked closely with trained service users/carers who:

- were involved in the design of the research studies in their role as co-applicants
- conducted data collection through the undertaking of in-depth interviews and the running of focus groups
- contributed as part of the research team to the coding and analytical frameworks developed as part of the research
- led peer-review publications disseminating research findings (Grundy et al., 2016; Cree et al., 2015)

Types of qualitative research

In this chapter we will focus on the qualitative data methods utilised within EQUIP. These are the main types of qualitative methods that researchers are likely to find themselves using in health services research:

- In-depth interviews
- Focus groups
- Observations

Service-user and carer involvement in qualitative research methods

Interviews

Qualitative in-depth interviews are useful for illuminating a range of perspectives from different types of participants. The types of qualitative interviews are illustrated in Figure 19.

Figure 19 **Types of qualitative interview**

Semi-structured interviews are the most commonly used type of interview and were used frequently in the EQUIP programme. By their nature, semi-structured interviews are less confined than more formally structured ones and require more improvisation on the part of the interviewer, but they also allow pre-set questions relating to the research aims and objectives to be explored. Semi-structured interviews still require significant preparation to be undertaken prior to the interview in the form of drafting of interview schedules. Care should be taken to ensure that the language used is simple and easy to understand to reduce ambiguities and that the questions and prompts included are sufficient to address the research aims and objectives. The schedule is usually refined during the course of data collection to explore emerging themes in more depth during subsequent interviews.

There is a range of different types of questions that be can be utilised in semi-structured interviews which will elicit different responses from research participants (Figure 20). Interviews will generally involve a mixture of these different types of questions and will be used within schedules to address specific research aims and objectives.

Figure 20 **Types of interview questions with examples**

Open/closed questions
- Closed question: Do you currently give service users a copy of their care plan?
- Open question: What do you think the benefits are of involving service users and carers in the care planning process?

Behaviour or experience
- Can you give me an example from your experience of where involving service users and/or carers in the care planning process has worked well?

Feelings
- How did your experience in a recent care-planning meeting make you feel?

Knowledge
- What are the current drivers of service-user/carer involvement in mental health care planning within your Trust?

Prompts
- Can you tell me a little bit more about that?
- How does that relate to...?
- Why do you think that might be?

Background demographic
– used to provide contextual information for the data presented (e.g. age, role, sex, diagnosis)

Interviews are most often carried out in person in a one-to-one format. For practical reasons, however, or if it is the preference of the participant, interviews can also be carried out over the phone or via secure Internet platforms. Interviews should be offered at a location and time that is convenient for the participant, although it is preferable for interviews to take place in a quiet location, free from interruptions and background noise. Getting the interviews, contents, structure and relationships right within qualitative interviews is critical to the success of any qualitative research project.

There are various strategies researchers can employ to increase the quality of data collected during interviews (Figure 21). Careful consideration should be given to the design of the study prior to commencing the project and sufficient flexibility should be factored into the design to allow for the iterative approach inherent in qualitative research methods.

Figure 21 **Interview strategies**

Funnelling
start with a broad, open question and then focus down

Story-telling
put the researcher in the role of active listener, rather than questioner

Reassuring
provide visual cues such as eye contact, head nods and supportive responses

Probing
further explore issues raised by participants with insightful follow-up questions

Linking
make links between a current comment and earlier ones

Naïve positioning
get clarification when something doesn't make sense

Acknowledging/ understanding
show your growing understanding and position yourself as a 'learner' and not an 'expert'.

Qualitative interviewing exercise

Develop an interview guide on the topic of mental health and wellbeing. Conduct the interview with a willing friend or colleague. Your interview guide can be written as an aid to hold a 'conversation with a purpose'. You want to understand what mental health and wellbeing might mean from an insider perspective. Write down the key points arising during the interview. You might want to consider and reflect on:

- the challenges of defining wellbeing:
  - What is it?
  - Could it be viewed differently from diverse points of view?
  - How would you find out about conceptualising mental wellbeing?

- where and when the interview should be carried out

- how did you find the process of undertaking an interview:
  - What types of questions did you use and what responses did these elicit?
  - Did you utilise any of the strategies for interviews detailed in Figure 21?
  - How might you change your approach in future interviews?

Focus groups

Focus groups are a form of qualitative interview in which a group of people, often with similar experiences, are brought together to speak about a given topic. The rationale for undertaking a focus group is that as a group, participants may generate data that they may not have during a one-to-one interview by virtue of the fact that they are discussing the topic with other people in a social interaction. The use of focus groups as a sole method of data collection has been discouraged because people may be uncomfortable discussing certain things in front of others (Mitchell, 1999). Therefore interviews and focus groups are often used in conjunction with each other, with participants being given the option of which they would prefer to attend (Cree et al., 2015; Grundy et al., 2016; Bee et al., 2015).

Observation

Observations are periods of intensive social interaction with people in their environment and are defined as 'the systematic description of events, behaviours, and artefacts in the social setting chosen for study' (Marshall and Rossman, 1989, p 79). Observations allow researchers to explore events that participants may not wish or may not be able to express during in-depth interviews. To qualify as research, observation should:

1. serve a focussed research purpose
2. relate to existing literature or theories
3. be systematically planned
4. be recorded systematically
5. be refined into general propositions or hypotheses
6. be subject to checks and controls on validity and reliability

During observations, researchers take field notes on participants, the setting, the social behaviour and the frequency and duration of events. They may also access other sources of data such as minutes of meetings observed or field interviews. The stages in observational work are illustrated in Figure 22.

In EQUIP, observations were used during the process evaluation as part of the trial to observe how patients and staff adopted and used the new user/carer involved care planning intervention. Observation included attention to how the new system fitted into the everyday routines of management and care practices for patients and professionals, and how data collected through the one-to-one interviews was supplemented.

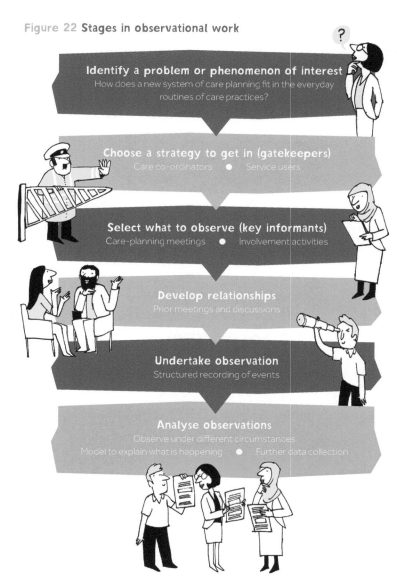

Figure 22 **Stages in observational work**

Exercise in observation

- Think about a topic that would warrant using observational methods rather than say a questionnaire or interview.
- Consider your reasons for this choice and why other methods would be less suitable.
- How do you think you could negotiate access into the setting related to your chosen topic?
- Who do you think would be the main focus of your observation?
- What might you want to record during the observation?

Qualitative sampling

Sampling refers to the process of selecting participants for a research study from the wider population. Qualitative sampling is purposive in nature and unlike quantitative sampling does not try to be representative. Instead it focuses on a depth of understanding and the principle that the smaller the number of participants a study has, the deeper the engagement with the participant and the greater level of understanding will be developed about the phenomena under consideration. Three common types of sampling are described in Figure 23.

Figure 23 **Commonly used types of sampling in qualitative research**

Purposive sampling
the most commonly used approach which group participants in line with predefined criteria relevant to the research question (e.g. service users under the care of a community mental health team)

Quota sampling
form of purposive sampling where it is decided in advance how many people with different characteristics will be recruited to the study and recruitment continues until the quota has been reached.

Snowball sampling
form of purposive sampling where existing contacts or participants identify other potential participants. Particularly useful for identifying hard-to-reach populations.

Saturation can help you know when to stop collecting data. Data saturation is reached when it is agreed amongst the research team that no new themes are arising from the data.

PPI stories from EQUIP

Next, Lindsey is going to describe her experiences of taking part in qualitative research as part of EQUIP.

LINDSEY'S STORY

I was a member of the EQUIP research team and was trained to conduct qualitative interviews and focus groups. I worked with colleagues to code and analyse the data we collected.

I really felt that my background as a carer helped me to understand and identify with many of the issues and setbacks that our research participants had experienced with mental health services. I had to work hard not to let my own views colour their opinions and perspectives.

This is an important skill that every qualitative research needs to develop. The themes that came out of our data were all too familiar to me and sometimes I'd come away feeling frustrated at not being able to wave a magic wand.

Running the focus groups was different. There were so many opinions all similar in nature, but there much less time to go into every one in depth. Everybody there wanted and needed to have their say. I learnt to manage this and had to wear a number of different hats. I was an observer, a notetaker and interviewer. Being able to stay empathic and professional, and being able to listen without jumping on board in agreement with someone else's story, was a big learning curve.

REFLECTIVE EXERCISE

- What is qualitative research and when should it be used?
- List three types of qualitative research methods.
- List 3 strategies to be employed during qualitative interviewing.

REFERENCES AND FURTHER READING

Brooks, H. L., Rogers, A., Sanders C. and Pilgrim, D. (2014) Perceptions of recovery and prognosis from long-term conditions: The relevance of hope and imagined futures. Chronic Illness. DOI:10.1177/1742395314534275.

Brooks, H. L., Sanders, C., Lovell, K., Fraser, C. and Rogers, A. (2015) Re-inventing care planning in mental health: stakeholder accounts of the imagined implementation of a user/carer involved intervention. BMC Health Services Research. 15:1, 490.

Cree, L., Brooks, H.L., Berzins, K. and Bee, P. (2015) Carers' experiences of involvement in care planning: A qualitative exploration of the facilitators and barriers to engagement with mental health services. BMC Psychiatry 15:1.

Grundy, A. C., Bee, P., Meade, O., Callaghan, P., Beatty, S., Olleveant, N. and Lovell, K. (2016) Bringing meaning to user involvement in mental health care planning: a qualitative exploration of service user perspectives', Journal of Psychiatric and Mental Health Nursing 23, 12–21. doi: 10.1111/jpm.12275.

Marshall, C. and Rossman, G.B. (1989) Designing qualitative research. Newbury Park, CA: Sage.

Michell, L. (1999) Combining focus groups and interviews: telling how it is; telling how it feels', in R. S. Barbour and J. Kitzinger (Eds),. Developing Focus Group Research: Politics, Theory and Practice. Thousand Oaks, CA: Sage Publications Ltd, pp. 36–46.

Ritchie, J. and Lewis, J. (Eds) (2003) Qualitative Research Practice: A Guide for Social Science Students and Researchers. Sage: London.

Chapter 8:
Introduction to Qualitative Data Analysis

Helen Brooks, Penny Bee and Anne Rogers

Chapter Overview

Qualitative data includes a range of textual (e.g. transcripts of interviews and focus groups) and visual (photographic and video) data. During qualitative analysis researchers make sense of this data gathered from research. Analysing the data by looking for common themes (known as thematic analysis) is one of the most common ways in which to do this and involves examining and recording patterns within the data relating to a specific research question. There are various criticisms levelled at qualitative analysis including issues relating to validity, reliability and credibility. Researchers can address these through a range of methods including triangulation of data, member validation, careful sampling and transparency of approach. The themes resulting from this form of analysis can illuminate participants' meanings, actions and social contexts relating to the phenomena under consideration.

Learning Objectives

By the end of this chapter you should be able to:

1. Understand what qualitative data is and how it can be analysed

2. Understand the factors that influence the choice of appropriate qualitative analysis methods

3. Understand how to carry out thematic qualitative analysis on qualitative research data

INTRODUCTION

The purpose of qualitative data analysis is to make sense of textual (e.g. transcripts from interviews and focus groups) and visual data (photographic and video) gathered through qualitative methods by identifying patterns and drawing inferences from them.

Qualitative data analysis can be:

- Inductive – analysis is guided only by the data collected during the study (e.g. akin to grounded theory)
- Deductive – analysis is guided by existing theories and frameworks (e.g. certain types of framework analysis)
- A combination of deductive and inductive approaches

Inductive approaches are most commonly used within qualitative data analysis with knowledge built from the ground up. Data analysis is usually carried out concurrently with data collection in line with the constant comparison method (Charmaz, 1995) so that any issues that emerge from the data can be explored in an iterative manner during future data collection and analyses (Figure 24).

Figure 24 **Constant comparison method of qualitative data analysis**

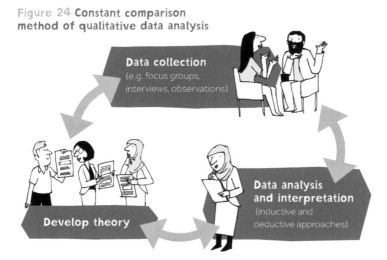

Thematic analysis

Thematic analysis is a core method for qualitative research and a flexible research tool which provides a rich and detailed account of data collected (Braun and Clarke, 2006). In a thematic analysis, researchers firstly read and re-read transcripts to ensure they are familiar with the data. Data is then searched to identify any recurrent patterns, which are then coded as such by the researchers (Figure 25). During the process of coding utilising a constant comparative method, researchers should actively consider the following questions:

- What is happening here?
- Under what circumstances does this happen?
- What is this data relating to?
- Are there any pre-existing codes this relates to? (Glaser, 1978)

> **Figure 25 Types of qualitative coding**
>
> **Open coding** – codes can be definitions, perspectives, processes or events
> - What does 'involvement' mean?
> - Who is care planning for?
> - Care planning meetings
> - Barriers to involvement
>
> **Axial coding** – the relationship between codes
> - Pre-requisite code
> - Outcome code
> - How one code impacts on another
>
> **Selective coding** – having established a code, looking for more examples to explore in more depth

Researchers then develop overarching themes from the codes. A theme is a cluster of linked categories conveying similar meanings, which emerge through the inductive analytical process, characterising the qualitative paradigm. Researchers should break off frequently from the process of analysis to write memos detailing their thoughts on particular themes and to reflect on any issues arising during the analytical process. Final themes are presented with supporting quotations from the raw data, and often brief detail on the demographic characteristics of the participant is also provided for contextual reference (Figure 26).

Figure 26 **Examples of themes from an EQUIP paper exploring the role of carers within the care planning process**

The structure and purpose of optimal care planning

What good care planning would be like? For us all to sit down and to build a picture of what my son would like to be doing in six months' time and how he would like to get there. And for us all to have a written copy of how that's going to happen and somebody to follow it through every stage of the way. Carer 1001, female, 53, cares for a son with a diagnosis of schizophrenia

Relational barriers to involvement in care planning

And I think there's an awful lot of... us and them, and a bit kind of pat you on the head, you're not expected to know what all this jargon means. Carer 1013, male, 27, cares for a brother with a diagnosis of bipolar disorder.

Confidentiality as a barrier to involving carers in mental health care planning

So it was like huge barricades up around this trivial information, trivial stuff. So... that in itself as you can imagine, was intensely upsetting and, and infuriating. But it's more that it symbolises this idea that as the carers you're nobody. Carer 1015 male, 45, cares for wife with a diagnosis of borderline personality

The analysis process should be undertaken by multiple researchers who code data independently. Researchers should meet regularly to discuss emergent analysis and to develop an agreement upon a set of codes. During these discussions specific consideration should be given to:

- alternative explanations of interpretations
- duplication of codes
- relationships between codes
- disagreement between researchers
- avenues for further exploration

These on-going discussions ensure that codes and resultant themes remain grounded in the data for purposes of validity. The constant comparison of new data means that the thematic framework can be amended and developed over time to allow for new codes to be introduced or redundant codes to be removed.

Thematic analysis exercise

Consider the content of the interview you conducted for the exercise in Chapter 7.

- What were the key themes to emerge from the interview?
  - What open codes are these based on?
  - How do these codes relate to each other?
- Did you identify any codes/themes that merit further exploration in any future interviews?

Ask a willing friend/colleague to independently identify the main themes emerging from the interview. Consider:

- how your emergent themes are similar and how do they differ?
- how you now feel about your original analysis?
- identifying three themes that you both agree on.

Triangulation of data

In addition to transcripts from interviews, qualitative data can also include observational data, diaries, photographs, digital forum discussions, social media posts and video recordings. For example, when interviewing participants about their experience of living with a chronic condition, you may want to ask them to capture their experience using photographs. These can be treated as a unit of data and analysed thematically in addition to the transcripts from interviews, and presented to support interpretations. (Figure 27)

Figure 27 **Example of photographic data presenting diabetes recording equipment**

By combining different types of data, you can add strength to your analyses and address some of the criticisms directed at one particular type of method. For example, in the case of interviews and focus groups it has been asserted that people may give socially acceptable accounts (public accounts) in formal research interviews that do not reflect their actual views and experiences (private accounts). By adding observations to the methodological approach, these concerns can be reduced.

Adding depth of understanding to randomised control trials (RCT)

Qualitative analysis can be useful when trying to understand why participants do, or do not, engage with interventions being tested as part of a randomised control trial (RCT). During EQUIP, this was explored using a longitudinal, qualitative process evaluation which ran alongside the RCT designed to test the training intervention. This involved:

- Semi-structured interviews – service users, carers and professionals sampled from both the intervention and control arm of the trial took part in three semi-structured interviews over the course of one year
- Observation of how service users and professionals adopt and use the new user/carer involved care planning
- Diary records of user and carer experiences of care planning

Analysis of this data identified a range of barriers to the use of the new user/carer involved care planning approach within mental health services which would not have been identified through the RCT alone. Examples included:

- Professionals cited time as a major barrier to involving service users and carers in care planning. A lack of resources within services meant caseloads were increasing and staff had limited time to spend with service users.
  - *"There's that pressure. People aren't being replaced. Erm...you know, people just expected to absorb more cases."*
- Service users acknowledged this lack of resources and described feeling under pressure to be discharged from services and minimal contact with their care team. Care co-ordinators were replaced frequently, meaning there was little time to build up relationships they considered as prerequisites to suitable involvement in the care planning process.

Trustworthiness of data and analysis

As with other types of data analysis, there are no strategies that guarantee trustworthiness of data, and the choice of how best to deal with issues of validity and reliability are normally at the discretion of the individual researcher. There are a number of issues to bear in mind when designing and undertaking qualitative analysis.

Validity

The validity of qualitative data refers to the 'trustworthiness' of the data or its ability to reflect the reality it is seeking to explore. Scientific validity is traditionally tested through replication. However, this is not possible with qualitative research due to the specific, context-dependent nature of the study design. Instead, careful attention is given to the context (both individual and societal) in which interviews are based and to the researchers carrying out the study. For example, the team should reflect on and make explicit any assumptions or bias they may bring to interviews. This can include theoretical positioning or any past experience that may have relevance. Here is an example of a reflexive statement about researcher positioning (Brooks et al., 2016):

HB and KR are health service researchers, SW is a Lecturer in Mental Health, KL is a Professor in Mental Health and AR is a Professor of Health Systems Implementation. As such, researchers had no therapeutic relationship with participants. The conceptual starting point of our study is one informed by a capabilities approach which recognises that social context and engagement with valued people, places and activities are often hidden from view but are likely to be as important to the management of long-term conditions as traditional therapeutic or self-management support approaches.

Researchers may also consider triangulation of data (discussed previously) and/or member validation. Member validation is commonly used to validate qualitative research findings. At its most basic level, it involves showing interviewees details of the analysis or summaries in order for them to confirm interpretations. During EQUIP, service users and carers were included as part of the analytical team and often led the analytical process to ensure that the data and any subsequent analysis reflected the reality of mental health care planning (Cree et al., 2015; Grundy et al., 2016). The production of a 'paper trail' (field notes and versions of coding frameworks) should also enhance the trustworthiness of qualitative data analysis.

Reliability and generalizability

Reliability refers to the ability of data to be consistent across time and contextual variations. In the natural sciences, this is argued with the defence of statistical significance and power calculations. In qualitative research, arguments are instead made for commonality or typicality (Fossey et al., 2002). Within the studies included in EQUIP, commonality was inferred by the fact participants were recruited by virtue of having certain characteristics (e.g. service users under the care of a community mental health team).

Credibility

Given the interpretative nature of qualitative analysis, credibility refers to the interpretations made about the data (Green and Thorogood, 2005). In order to address credibility, emerging themes should be discussed and tested with the wider research team to ensure concepts and themes derived from the data are rooted in the raw data itself.

PPI stories from EQUIP

Next, Lindsey is going to describe her experiences of undertaking qualitative data analysis as part of the EQUIP project.

LINDSEY'S STORY

During EQUIP I worked with experienced colleagues to analyse a lot of qualitative data. Gathering and analysing the data and putting them into themes was really exciting, but also daunting. I suddenly realised the number of issues that had been raised during our interviews and focus groups, that they were real issues for people and that we needed to look at them in depth. I remember feeling really inspired when the need to train professionals was identified as a recurrent topic. I was also heartened that issues of confidentiality, which had raised a lot of concern with carers, were going to be taken seriously by our research team.

I was the lead author on a paper which set out to document the experiences of carers in mental health services, which was published in BMC Psychiatry. I still pinch myself occasionally. In my world, research papers have always been written by professionals. My work on EQUIP has made a difference but I couldn't have done it on my own. The research team have been supportive and never once made me feel inept. Considering I left school with no qualifications, our achievements have been amazing.

REFLECTIVE EXERCISE

- Describe the two main approaches to qualitative analysis.
- Describe and outline the main stages of thematic analysis.
- What are the main criticisms of qualitative data analysis and what strategies can researchers employ to overcome them?

ALLIED EQUIP PAPERS

Bee, P., Brooks, H.L., Fraser, C. and Lovell, K. (2015) Professional perspectives on service user and carer involvement in mental health care planning: A qualitative study. International Journal of Nursing Studies, 52:12, 1834–1845.

Brooks, H. L., Rushton, K., Walker S. and Rogers, A. (2016) Ontological security and connectivity provided by pets: A study in the self-management of the everyday lives of people diagnosed with a long-term mental health condition. BMC Psychiatry, 16, 409. https://doi.org/10.1186/s12888-016-1111-3

Brooks, H. L., Sanders, C., Lovell, K., Fraser, C. and Rogers, A. (2015) Re-inventing care planning in mental health: stakeholder accounts of the imagined implementation of a user/carer involved intervention. BMC Health Services Research, 15:1, 490. https://doi.org/10.1186/s12913-015-1154-z

Cree, L., Brooks, H. L., Berzins, K. and Bee, P. (2015) Carers' experiences of involvement in care planning: A qualitative exploration of the facilitators and barriers to engagement with mental health services. BMC Psychiatry, 15:1. https://doi.org/10.1186/s12888-015-0590-y

Grundy, A. C., Bee, P., Meade, O., Callaghan, P., Beatty, S., Olleveant, N. and Lovell K., (2016) Bringing meaning to user involvement in mental health care planning: a qualitative exploration of service user perspectives. Journal of Psychiatric and Mental Health Nursing, 23, 12–21. doi: 10.1111/jpm.12275

REFERENCES AND FURTHER READING

Braun, V. and Clarke, V. (2006) Using thematic analysis in psychology., Qualitative Research in Psychology, 3:2, 77–101.

Charmaz, K. (1995) 'Grounded theory', in Smith, J. A., Harre, R. and Van Lagenhave, L. (Eds), Rethinking methods in psychology. London: Sage, pp. 27–49.

Cornwell, J. (1984) Hard-earned lives: Accounts of health and illness from east London. London: Tavistock.

Glaser, B. G. (1978) Theoretical sensitivity. California, The Sociology Press.

Green, J. and Thorogood, N. (2005) Qualitative methods for health research. London: Sage.

Fossey, E., Harvey, C., McDermott , F. and Davidson L. (2002) Understanding and evaluating qualitative research. Australian and New Zealand Journal of Psychiatry, 36: 717–732.

Strauss, A. and Corbin, J. (1990) Basics of qualitative research: Grounded theory procedures and techniques. Newbury Park: Sage.

CHAPTER 9:
PRINCIPLES OF ETHICAL RESEARCH

Owen Price and Lauren Walker

CHAPTER OVERVIEW

By definition, research seeks to explore something that is unknown. This uncertainty means there is always the possibility of harm arising from research. There are many examples in both near and distant history of serious harm to participants as a consequence of research, including permanent disability and death. This is why it is of great importance that research projects are informed by sound ethics, properly planned, approved by an independent ethical board and rigorously monitored throughout the duration of the study. This chapter will introduce four principles that govern the conduct of ethical research using relevant case examples to bring each principle to life.
Topics explored in the chapter include:

- The importance of 'informed consent'
- Assessment of capacity to provide consent
- Measures to minimise and manage harm arising from planned research
- The importance of ensuring that possible benefits of the research outweigh the risks and costs to participants
- Ensuring that participants are treated fairly and equally throughout the study

LEARNING OBJECTIVES

By the end of this chapter you should be able to:

1. Explain why sound ethical procedures for the planning and conduct of research are important

2. Understand key principles that govern the conduct of ethical research

3. Illuminate each principle with case examples relevant to service-user-led research.

INTRODUCTION

Following the end of World War Two, sixteen Nazi doctors were tried and convicted of war crimes involving research on humans (Seidelman, 1996). During this research, people were forced to participate in 'medical' procedures that involved torture and often resulted in death (Weindling, 2005). During the trials, the doctors' defence lawyers argued that there were no existing laws that clearly defined the difference between legal and illegal research involving humans (Annas, 1992). Following the trials, the 'Nuremberg Code' and 'Declaration of Helsinki' established a set of ethical rules for research involving humans, designed to prevent similar atrocities from happening again. These rules related to four broad ethical principles that govern the conduct of medical research. These are:

- **Autonomy** – this means the right of people to make their own decisions

- **Non-maleficence** – this means doing no harm

- **Beneficence** – this means acting in people's best interests

- **Justice** – this means treating all people fairly and equally (Beauchamp and Childress, 2001).

Despite these principles, tragedies involving research continue to occur. You may recall the 'elephant man trial' (so-called because of the deformities caused by the trial medicine TGN1412) at Northwick Park Hospital in 2006. Six healthy volunteers ended up fighting for their lives in intensive care after being administered with the trial medicine. The researchers had failed to adequately consider, before trialling the medicine, what a safe dose for humans was likely to be (HMSO, 2006). This demonstrates how important it is that the process of approving and monitoring research studies is highly rigorous. This chapter will explain each of the four ethical principles above, using case examples to bring them to life.

Ethical principles

AUTONOMY – right to choose

Autonomy refers to the ethical duty of researchers to take active steps to ensure the person makes an independent decision about whether or not to take part in a research study. Anyone who takes part in a research study must provide informed consent. This means that people must give their permission to participate, with full understanding of what they are consenting to, free from pressure from others. A person cannot provide informed consent if they lack capacity. A person may be considered to have capacity if they can understand information provided about the study and the advantages and disadvantages of taking part. Importantly, they must also be able to form an independent decision on this and be able to communicate their decision to researchers.

Discussion point 1:
Considering the above, read **case example A** in **Figure 28**.

Figure 28 Case example A

You are a service user researcher doing an interview study on a mental health ward. You are recruiting in-patient service users to explore their experiences of physical restraint. A service user, Andrea, approaches you and says she is keen to take part. You speak to the nurse in charge and say that you would like to speak with Andrea and provide more information about the study. The nurse in charge informs you that, in her view, Andrea is not currently well enough to understand the risks and benefits of taking part and that, furthermore, the topic of the interviews is likely to cause significant distress to Andrea.

Would you:

a) Ignore the nurse's advice. It's Andrea's decision whether or not to take part.

b) Speak to Andrea, apologise, and say that it is not possible for her to take part.

c) Leave the study information with Andrea, apologise and say that she is not able to take part at the present time because the nurses don't think she is well enough. Assure Andrea that you will continue to contact the ward to find out when the situation has changed.

The Mental Capacity Act (2005) requires that a person with a 'duty-of-care' (in this instance, the nurse in charge) should make a decision where a person may lack capacity. However, no assumptions should be made about the person's capacity based on factors not related to the specific task (i.e. in **case example A** this would be whether or not the service user had the ability to understand what the interview would involve and the risks and benefits of taking part).

Any decision must also acknowledge that capacity can change over time. It is also worth considering the potential vested interest the nurse has in preventing service users from reporting negative care experiences given the research topic in **case example A**. For these reasons, answer c is correct because the researcher has complied with the legal requirement to assess capacity but has also protected Andrea's right to participate in the study in the future.

To provide informed consent, a potential participant must have access to complete information about the study, including what their participation will involve. This information should include the aims, methods, risks and benefits of taking part and their right to withdraw from the study at any stage without penalty (Beauchamp and Childress, 2001). **Figure 29** provides a list of the key ingredients you should expect to find on a study information leaflet to ensure the potential participant is making a fully informed decision.

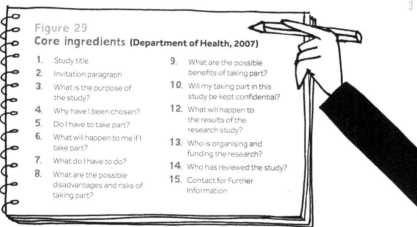

Figure 29
Core ingredients (Department of Health, 2007)
1. Study title
2. Invitation paragraph
3. What is the purpose of the study?
4. Why have I been chosen?
5. Do I have to take part?
6. What will happen to me if I take part?
7. What do I have to do?
8. What are the possible disadvantages and risks of taking part?
9. What are the possible benefits of taking part?
10. Will my taking part in this study be kept confidential?
12. What will happen to the results of the research study?
13. Who is organising and funding the research?
14. Who has reviewed the study?
15. Contact for Further Information

NON-MALEFICENCE – do no harm

Non-maleficence refers to the ethical responsibility not to harm the participant during the research. Obvious examples of potential harms from research include death, pain, injury, distress, offence or neglect. However, there is also a duty to avoid less obvious forms of harm such as to treat unfairly or act against participant interests, including wasting their time (Beauchamp and Childress, 2001).

Discussion point 2:

Consider **case example B** in **Figure 30**
– what harms have arisen from the research?

Figure 30 Case example B

You are a service user researcher doing an interview study on a mental health ward exploring in-patient service user experiences of physical restraint. You have recruited a service user, John, who has been assessed as having capacity, has had 48 hours to read the study information and has provided fully informed consent. Part way through discussing a restraint experience, John becomes extremely tearful and anxious and tells you he plans to harm himself after the interview.

Clearly the research topic in **case example B** had the potential to cause distress to participants. A study involving in-patient mental health service users is unlikely to receive ethical approval without a procedure for informing professionals if evidence they are at risk emerges during their participation. This means that John was also likely to be exposed to a second harm, having his confidentiality breached.

Any study that has the potential to cause distress should have an approved protocol to minimise and manage distress should it arise. This should include the nature of the support to be provided both during and after participation, emphasising that participants can end participation without penalty and have their data destroyed. Importantly, the wellbeing of the participant must be prioritised over the scientific benefit of their continued participation (Mental Capacity Act, 2005). Where there is a risk of breaking participant confidentiality, what participants say that could lead to their confidentiality being broken should be made clear to them verbally and in the written study information before participation.

BENEFICENCE – do good

For a research study to be considered ethical, it must prove it has enough potential benefit to warrant the time and other costs participants give up to take part. Importantly, the possible benefits to participants can be indirect (i.e. they can be of benefit to the participant through benefiting society or a wider social group). Crucially, the potential benefits must outweigh the potential costs to participants.

Discussion point 3: Consider **case examples A** and **B**. What are the potential benefits to service users and society of service user participation in this study? How do you think these weigh up against the possible harms identified in the previous section?

The study may have a number of benefits. Physical restraint can cause serious physical and mental harm to service users and staff (Bonner et al., 2002; Paterson et al., 2003; Renwick et al., 2016). Use of physical restraint is also expensive for the NHS (Flood et al., 2008). Research that draws focus on the issue and that may lead to policy changes reducing restraint use could therefore have possible benefits to both service users and society. When weighing up risks against benefits, it is important that the risks are not overstated and participants are not denied the benefits of making a contribution.

JUSTICE – treat fairly and equally

Researchers must treat participants and potential participants fairly and equally in order for a study to be considered ethical. This does not mean that all participants must be treated exactly the same way during a research study or a randomised controlled trial. Chapter 3 provides a clear example of when this is not the case. However, sometimes researchers will be expected to provide an effective trial intervention to participants in the control group after a study has finished. This is because of the need to treat participants fairly and equally.

Researchers must consider fairness and equality when considering who to include in their study. Again, this does not mean it is ethically unacceptable to exclude people or groups on the basis of certain characteristics that would make their contribution to achieving the study's aims irrelevant. On the other hand, it does mean that every effort should be made to ensure those with a relevant contribution to make should be given fair opportunity to participate. Once participants have been recruited, the obligation to treat them fairly does not end. Researchers must consider how they are going to compensate participants for the time and other expenses they sacrifice to take part. Such payment should reflect the contribution they make. However, any payment should not represent such an incentive that the person feels unduly influenced to take part because they do not want to miss out on the payment. This would represent a violation of their right to choose (autonomy) free from the influence of others (Beauchamp and Childress, 2001).

This chapter has examined how issues of capacity and vested interest can potentially interfere with mental health service user participation in research. It provides a useful example of how and when the four ethical principles overlap and interact. Consider **case example A**: the nurse may have refused to allow the service user to participate out of a reasonable expectation that the research would harm the participant (Beauchamp and Childress, 2001).

If this assessment of potential harm was accurate, then this may have been an occasion where it was appropriate, in the short term, to deny the potential participant the right to choose (autonomy) and the benefits of participation (beneficence). However, if vested interest was at play, and the nurse was trying to silence critical views of care experiences, this would violate the participant's right to choose, the benefits of taking part, their right to be treated equally and fairly and, as such, would expose them to harm. This is why it is so important that, where a decision is taken to deny a participant the right to choose, this has been based on a sound, justifiable decision that has adequately balanced the potential risks of participation against the potential benefits. Furthermore, that a proper plan has been put in place to restore the potential participant's right to choose at the earliest stage possible (i.e. ensuring the person is able to participate when their circumstances have changed).

REFLECTIVE EXERCISE

- What are the four key features of ethical research?
- Give three examples of how a research project could do harm to a study participant.
- Describe three ways in which you can your protect the confidentiality of research.

REFERENCES AND FURTHER READING

Annas, G. (1992) The Nazi Doctors and the Nuremberg Code. New York: Oxford University Press.

Beauchamp, T., and Childress, J. (2001) Principles of Biomedical Ethics (5th Ed). New York: Oxford University Press.

Bonner, G., Lowe, T., Rawcliffe, D., and Wellman, N. (2002). Trauma for all: a pilot study of the subjective experience of physical restraint for mental health inpatients and staff in the UK. Journal of Psychiatric & Mental Health Nursing, 9(:4), 465-73. Epub 2002/08/08.

Department of Health (2007) Information sheets and consent forms guidance for researchers and reviewers, version 3.1. London: DH.

Flood, C., Bowers, L., and Parkin, D. (2008) Estimating the costs of conflict and containment on adult acute inpatient psychiatric wards. Nursing Economics, 26(:5), 325-30.

HMSO (2006) Expert scientific group on phase one clinical trials. Final report. London: HMSO.

Mental Capacity Act (2005) Mental Capacity Act, 2005. (c.9) London: HMSO.

Paterson, B., Bradley, P., Stark, C., Saddler, D., Leadbetter, D., and Allen, D. (2008) Deaths associated with restraint use in health and social care in the UK. The results of a preliminary survey. Journal of Psychiatric & Mental Health Nursing, 10:1,3-15.

Renwick, L., Lavelle, M., Brennan, G., Stewart, D., James, K., Richardson, M., et al. (2016) Physical injury and workplace assault in UK mental health trusts: an analysis of formal reports. International Journal of Mental Health Nursing, 25(:4), 355-366, doi: 10.1111/inm.12201.

Seidelman, W. (1996) Nuremberg lamentation: for the forgotten victims of medical science. British Medical Journal, 313:,1463.

Weindling, P. (2005) Nazi Medicine and the Nuremberg Trials: From Medical War Crimes to Informed Consent. London: Palgrave Macmillan.

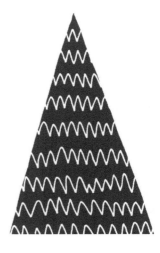

CHAPTER 10:
RESEARCH DISSEMINATION AND IMPACT

Helen Brooks and Penny Bee

CHAPTER OVERVIEW

Research activity does not finish when data analysis is complete. Once research findings are available, researchers still have obligations to fulfil. These obligations include sharing the findings with different audiences and ensuring maximum impact from the study.

The process of sharing research learning with others can be an enjoyable but challenging one. Often it is referred to as dissemination, but you may also see it linked with terms such as knowledge transfer or knowledge mobilisation. Each of these concepts is slightly different.

Dissemination refers to the active process of communicating research findings in a targeted and personalised way to identified relevant audiences who may be interested in the findings and/or able to benefit from them. Knowledge transfer extends beyond this dissemination phase and refers to an often lengthier process that includes both dissemination and the exchange and application of new knowledge in order to provide more effective health services and to strengthen health systems. In this chapter we will focus primarily, although not solely, on dissemination. There are ranges of ways in which research findings can be disseminated and some of these are discussed in the following pages.

LEARNING OBJECTIVES

By the end of this chapter you should be able to:

1. Explain the importance of disseminating research findings
2. Introduce different ways to disseminate research findings and increase research impact
3. Demonstrate the value of continuous stakeholder engagement for research dissemination.

INTRODUCTION

Mental health care resources are finite. In order to ensure service users receive the highest quality health care, evidence about the most effective and acceptable treatments needs to be fully incorporated into health care policy and practice. However, we have known for a long time that this is not happening as well as it should be within health services and that research evidence is not being transferred sufficiently to routine clinical practice both in the UK and across the world. This is often because of a failure by researchers to disseminate their work appropriately.

Dissemination can be defined as:

'A planned process that involves consideration of target audiences and the settings in which research findings are to be received and, where appropriate, communicating and interacting with wider policy and health service audiences in ways that will facilitate research uptake in decision-making and practice'. (Wilson et al., 2010).

The majority of research funders now require that applicants provide a dissemination strategy outlining the various ways in which the findings of the study will be disseminated to interested parties.

Key audiences implicated in this process usually include:

study participants

the public

health professionals

health care managers and policy makers

Commissioning organisations
(e.g. NHS England)

External Organisations
(e.g. NICE and the Department of Health)

Other researchers and academics

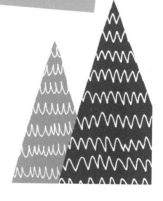

Each of these audiences may be interested in and require a different type or level of knowledge. For examples, study participants often want to know the results of the study to which they contributed their data. The general public may be less interested in the results per se and more concerned with what they mean for their own future health care decisions or care needs. Policy makers and service commissioners may be interested in what the results tell them about improving population health outcomes, as well as the cost of providing new or improved services to local and national communities. We know from recent evidence that the best way to disseminate research is to include a range of different proactive approaches which are appropriate and targeted to the different audiences with whom researchers need to engage. Ideally, dissemination activities should start early in the research process and include face-to-face interactions to maximise other people's engagement and interest in the study findings.

Research impact

Impact has been defined as the 'demonstrable contribution that excellent research makes to society and the economy' (ESRC, 2017). Impact can be achieved by influencing health policy, practice or behaviour or by building capacity amongst service users, carers, professionals or researchers. Funders often require researchers to define within applications for funding the pathways through which the study will demonstrate academic, economic and societal impact. In order for such impacts to be realised in practice, the findings from a research study must be disseminated suitably.

Types of dissemination activities

Researchers outline their plans for disseminating the results of a research study in a dissemination strategy. This should be co-produced with appropriate patient-and-public representatives as early as possible in the design of a research study to optimise the impact of the study and to ensure dissemination activities are suitable and appropriate for relevant target audiences. The selection of specific engagement activities and communication channels should be informed by current evidence on dissemination and knowledge mobilisation. For example, starting the process of engaging with relevant audiences early in the research process is more effective than waiting until the end of the project. It is very important to cost in dissemination activities (e.g. open access costs, travel for presentations, printing costs etc.) within the project budget to ensure the resources are available when they are needed.

Study participants

Providing study participants with the results from a research project in which they have been involved acknowledges their contributions and demonstrates the value their input has had. The most commonly used way to feed back to study participants is through a written lay summary of the research findings. Including an option on study consent forms allows participants to register their interest in finding out about the results of a study and register their details to receive written feedback. Service-user and carer representatives should be central to this process to ensure that the feedback is written in a comprehensive and easily accessible format.

For those studies not requiring consent from participants in a formal way (e.g. anonymous questionnaires), lay summaries may be disseminated in different ways. For example, an overview of the findings may be posted on study websites or social media pages. Participants can then be signposted to such locations by including a link within the questionnaire, along with the date when the lay summary will be made available. Hard copy lay summaries may be distributed through the organisations that have been involved in recruitment for the study, including for example healthcare trusts and/or local and national voluntary and community organisations.

Increasingly, researchers are also considering more interactive and innovative ways to disseminate findings to study participants if they have the budget and resources available. Possible examples include: infographics, interactive DVDs, video abstracts and animations of study findings, as well as having patient and public involvement representatives organise their own dissemination conferences and present study findings to interested groups.

Patients and the public

Researchers have a social obligation to disseminate findings to the general public especially if the topic is considered to be of significant relevance to wider society. The general public is likely to be current and future health service users, as well as relatives, family and friends of service users. As a result, in a dissemination strategy equal consideration should be given to disseminating findings to the general public as to publishing academic papers or conference proceedings.

Certain outputs, such as patient information cards, may usefully communicate study findings to current NHS service users, especially if they can be used to empower these individuals and stimulate demand for better quality, more effective healthcare.

The EQUIP study, for example, produced and disseminated a patient-mediated information card to help mental health service users input and shape their own care plans. This card was co-developed with service users and carers and designed specifically to communicate research findings in a way that met their needs.

There are many ways that public dissemination can be achieved, and researchers may also consider one or more of the following:

- Producing a press release for distribution to local and mainstream media outlets
- Presenting the study findings at local and national community events (e.g. service-user and carer forums or annual events such as the Mental Wealth Festival)
- Organising events such as pop-up dissemination cafes in local venues or arranging a conference where researchers and patient and public researchers engage with the public.

- Approaching well-known bloggers or vloggers in the field and asking if they are interested in writing an article about the research
- Developing a social media campaign led by service-user and carer researchers which is specifically targeted in terms of relevance and potential influence (including writing blogs, starting Facebook groups, organising TweetChats which allow Twitter users to participant in real-time hashtag conversations about the research or hosting a Reddit Ask Me Anything Session)

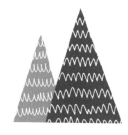

Figure 31 **Example impact from use of Reddit Ask Me Anything Session.**

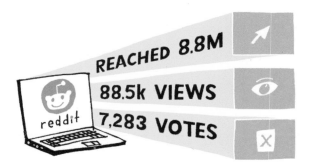

The research community

The most common route to reporting study findings to academic audiences is through publication in peer-reviewed scientific journals or through presentations at relevant national and international academic conferences. There are various guidelines available for researchers to support the process of reporting research to ensure that findings presented allow replication of the study and that they are presented in such a way as to allow the data to be included in future evidence reviews. These include the CONSORT guidelines for the reporting of randomised controlled trials and the COREQ guidelines for the reporting of qualitative research. Wherever possible, publications should be made 'open access' which means there are no restrictions (e.g. subscription licences) on who can view the articles. Open access articles can be found and read in full through a keyword internet search. Service-user and carer representatives should be included as co-authors when writing manuscripts for publication and invited as co-presenters at conferences.

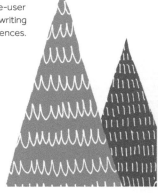

Healthcare professionals and relevant statutory and community organisations

In addition to the engagement strategies described above, researchers should also consider how they wish to engage with stakeholders within health-care services. When researchers are developing a dissemination strategy, they should work with the wider study team to draw up a list of potential health professionals, healthcare organisations, commissioning organisations and external voluntary or community organisations that may be interested in their study. This list can be updated over the course of the research to give an accurate, up-to-date overview of key audiences. Liaising with these stakeholders to discuss relevant dissemination techniques will ensure that the research team are using the most appropriate strategy for each organisation, to ensure maximum reach for their research findings. Activities specific to this audience may include producing articles for staff newsletters, presenting the findings as part of local or national seminar series, providing dissemination sessions to staff, managers or commissioners or attending healthcare events and activities to promote the study.

Tips to facilitate research dissemination

- Establish networks and relationships with service users, carers, professionals and organisations from the outset of the study
- Acknowledge the central role and importance that users and carers have in the process of dissemination and draw on their expertise
- Involve networks in all stages of research including dissemination activities
- Be flexible but consistent in your dissemination activities
- Understand the contexts in which you are undertaking dissemination and take the lead from your local networks
- Obtain management-level support from any organisations with whom you wish to engage
- Provide time and sufficient resources for dissemination activities within your project plan and budget

Knowledge mobilisation

Knowledge mobilisation extends beyond dissemination, and includes active efforts to change and influence practice. Evidence suggests that knowledge mobilisation to improve health care relies not only on producing new research outputs but also on brokering this knowledge to enhance the uptake of research evidence in situ. This can be assisted by first identifying and then overcoming potential boundaries and barriers to the flow of knowledge. Teams may need to invest time and effort in building trusted and enduring partnerships with services, service managers, commissioners and policy-makers.

For the most part, knowledge mobilisation is likely to be more successful if multiple activities and strategies are combined. Locally, the interpretation of research evidence may be challenged by entrenched professional identities and collective practice, and in such cases local opinion leaders and interactive educational meetings, facilitated by a mix of academic and service user/carer researchers, can be a powerful way of raising awareness and stimulating the momentum for practice change. To encourage wider roll-out of the outputs of research, teams may in addition consider holding a stakeholder conferences to engage regional or national audiences. In certain cases, they may also consider establishing online knowledge repositories with downloadable resources that link directly to patient and professional networks or organisations. Patient-and-public representatives can play a role in all of these activities, making sure that patient priorities are reflected in these communications, and that the right messages are given in the right way to the right people at the right time.

There is limited value in doing research unless you let people know about it. This chapter has examined the importance of disseminating research studies and their findings and considered the ways in which researchers can do this. Dissemination activities should target as wide an audience as possible using individualised strategies targeted to specific audiences, including drawing on the expertise of local collaborators and networks.

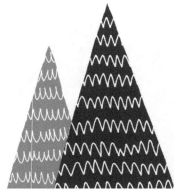

PPI stories from EQUIP

Next, Garry talks about his experiences of disseminating the research findings from EQUIP.

Garry's story

My involvement in research began when I was invited to participate in a focus group. Since then, I have been invited through my local networks to get involved in research myself. I have become a research project advisor and have felt a great sense of achievement.

It feels so rewarding that as members of project advisory panels, our feelings and opinions are listened to. I have inputted into a range of study outputs, including academic publications, animated resources, and written information and wallet cards for service users. I have also contributed to professional education and development events. I was recently invited to a research awareness event attended by student mental health professionals. I felt able to communicate with this group of learners the importance and value of PPI involvement in the research process, and this approach was well received. I've been surrounded by hard-working and supportive professionals for so long now, so it's great to meet the workforce of the future and engage with people who will influence care improvements in the world of mental health.

Reflective exercise

Amira is developing an application for a five-year programme of research designed to develop and test the effectiveness of a telephone support intervention for depression and anxiety. The research will also look at who is best placed to deliver this intervention (e.g. either within the NHS or within the voluntary sector) and what the most cost-effective option is.

- When should Amira start thinking about her dissemination strategy and who should be involved in developing it?
- Who do you think are the key audiences who might be interested in the research and its findings?
- What do you think are the most appropriate strategies for disseminating the findings to each audience and who is best placed to lead on each?
- How might Amira demonstrate the potential impact of her study to the funders?

REFERENCES AND FURTHER READING

Economic and Social Research Council (2017) What is impact? Available: http://www.esrc.ac.uk/research/impact-toolkit/what-is-impact/ [Accessed 3 October 2017].

Wison, P. M., Petticrew, M., Calnan, M.W., and Nazareth, I. (2010) Disseminating research findings: what should researchers do? A systematic scoping review of conceptual frameworks. Implementation Science, 5:91.

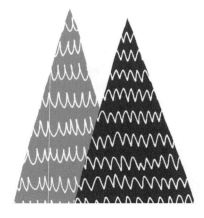

SUMMARY

This handbook has been written to help patient and public representatives engage in health services research and work meaningfully with academic and clinical research teams in true partnership. It has been co-written with service users and carers from the NIHR EQUIP research programme and aims to help other public and patient representatives increase their understanding and skills in research methods.

Health research is incredibly important. It helps to develop and evaluate new treatments, improve patient safety, and identify the most effective ways to organize, manage, finance, and deliver high quality care. As a member of the public, you will have your own health experiences and a unique viewpoint on the direction that this research could take.

We have integrated research methods training with personal stories and reflections from our PPI representatives throughout this book, and hope this has provided a useful resource.

As our stories have shown, there are many different roles and avenues through which members of the public can contribute to research. This means that different levels of involvement are possible and different people will have different amounts of time available and different preferences for what they would like to do.

Whatever your choice or intention, we wish you luck in your future research activities. We hope that this book has gone some way to equipping you with the knowledge and skills that you need to make a valuable and enjoyable contribution.

'I would just like to say thank you for giving me the opportunity to take part in the research programme; it made me feel so good about myself as it really boosted my confidence. It made me realise just how much of a difference I can make with the right tools and training.' (EQUIP project advisory group member)

'I learned so much by being able to take part in the research programme. It has been a dream come true attending university; it just proves that no matter what disabilities someone may have – physical or mental – with help, dreams can come true. I hope I can be part of future research programmes and I look forward to finding out what the future holds for me.' (EQUIP research methods course participant)

Glossary of key terms

Baseline
The starting point or measurement used for comparisons.

Bias
A type of error that can affect research and distort research findings.

Clinical significance
The practical significance of a result, or its significance in the real world.

Confounding variables
Extra variables that researchers may or may not have accounted for, which can alter the results of study or an experiment.

Cognitive functioning
Intellectual activities that that allow you to become aware of, perceive or understand ideas.

Cohort studies
Longitudinal studies that follow a group of people with a shared, defining characteristic over time.

Control/comparison group
The group of participants in a study that does not receive treatment or receives an alternative treatment and is compared to the group that receives a new intervention or treatment under test.

Constant comparison
An analytical process used in qualitative research to compare emerging results with new data as it is collected.

Content validity
The extent to which items on a measurement scale are judged to be relevant indicators of the variable being measured.

Confidence interval
The level of uncertainty around an estimated effect size (see below) reported in a research study.

Correlation
A statistical measure describing the strength of association between two variables.

Cross-sectional
A study that gives researchers a snapshot of a population at a single point in time.

Data synthesis
The process of combining results from different studies or from different sources of data within the same study.

Deductive analysis
A form of qualitative analysis that is guided by existing theories and ideas.

Dependent variable
The outcome variable expected to change as a result of a change in another variable.

Economic evaluation
A way of systematically identifying the costs and benefits of different health activities.

Effect size
Quantifies the difference between two or more groups differing in a given characteristic or level of treatment exposure.

Exclusion criteria
Characteristics that prohibit participants from taking part in research projects or disqualify studies from being included in systematic reviews.

Generalisability
The extent to which research findings can be transferred to other settings

Hypotheses
A set of predictions based on known facts that have not yet been tested and which act as a starting point for research.

Inclusion criteria
The characteristics that potential research participants must have to be eligible to take part in a study or the characteristics that potential studies must have to be included in a systematic review.

Incremental Cost-Effectiveness Ratio (ICER)
The difference in the costs of the two treatments divided by the difference in benefit.

Independent variable
A predictor variable that is being manipulated in an experiment

Inductive
A form of qualitative analysis that emerges only from the data collected during the study.

Informed consent
The process of giving permission to participate in something, with full understanding of what you are consenting to, and free from pressure from others.

Intention-to-treat analysis
The need to include all participants randomised in a trial in the final analysis of data, regardless of whether or not they recived their allocated treatment.

Inter-rater reliability
The relationship between the ratings of two independent raters scoring the same behaviour.

Internal consistency
How well individual items (or questions) in a scale measure the same thing.

Longitudinal
A study which collects research data at multiple time points.

Member validation
Showing interviewees details of a qualitative analysis or summary in order for them to confirm the data interpretation.

Meta-analysis
A method for combining data from multiple quantitative studies to produce a single conclusion with greater precision and statistical power.

Null hypothesis
The notion there will be no effect or relationship between two variables.

Odds ratio
The chance that a particular outcome will occur given a particular context or treatment.

P-value/Statistical significance
An interpretation of statistical data that tells us that an occurrence was most likely the result of a relationship between variables and not simply due to chance.

Population
A defined group of people with similar characteristics

Power calculation
Calculates the minimal sample size required to detect a significant difference between the treatment and control groups in a randomised trial.

Pre-post test design
Research in which participants are studied before and after an intervention.

Predictive validity
Occurs when people's scores on one scale predict other related things.

Primary outcome
An outcome the researchers consider to be of most importance within a study.

Protocol
A detailed document that describes the background, objectives, methodology, statistical plan and management of a study.

Psychometrics
The term used to describe the science of psychological testing and measurement scale development.

Purposive sampling
Participants are selected in line with predefined criteria relevant to the research question.

Quality-adjusted life-years (QALY)
A measure of both quantity of life gained, and the quality of the health that is achieved

Randomisation
A process of allocating participants to the treatment or control group by on the basis of a coin-toss.

Reliability
The degree of stability with which a scale measures what it is designed to measure, the likelihood that it will measure the same on two or more occasions.

Samples
Subsets of populations chosen to represent the bigger group.

Sampling
The process of selecting participants for a research study from the wider population

Saturation
The point at when there is nothing new arising from the qualitative data.

Secondary outcome
A outcome that researchers consider to be important in a research study but of secondary importance when compared to the primary outcome

Semi-structured interviews
Interviews consisting of a range or closed and open-ended questions but also allowing some freedom in terms of interview content

Standard deviation
The extent to which values in a dataset are clustered (or not clustered) around the mean.

Systematic review
A method used by researchers to identify, analyse and combine the findings of multiple studies in a rigorous way.

Randomised Controlled Trial (RCT)
A research project in which participants are randomly allocated to receive either an experimental or comparison intervention.

Themes
Patterns identified by researchers that occur in qualitative data and are related to the topic of interest.

Triangulation
The combining of different types of data to add strength and confidence to research findings or analyses.

Type 1 error (false positive)
When a researcher concludes that two variables are related to one another when they are not

Type 2 error (false negative)
When a researcher concludes that two variables are not related when in fact they are.

Validity
The degree to which a scale measures what it was designed to measure

Willingness to pay threshold (WTPT)
The maximum amount that a healthcare organisation is prepared to pay to improve the health of its community

Lightning Source UK Ltd.
Milton Keynes UK
UKHW020012300822
407907UK00006B/184